Gerhard Heyer
Hans Haugeneder (Eds.)

Language Engineering

Ralf Jungclaus
Modeling of Dynamic Object Systems

Christoph W. Keßler
Automatic Parallelization

Jürgen M. Schneider
Protocol-Engineering

Gerhard Heyer/Hans Haugeneder
Language Engineering

Dietrich W. R. Paulus/Joachim Hornegger
Pattern Recognition and Image Processing in C++

Dejan S. Milojicic
Load Distribution

Reinhard Gotzhein
Open Distributed Systems

Vieweg

Gerhard Heyer
Hans Haugeneder (Eds.)

Language Engineering

Essays in Theory and Practice of Applied Natural Language Computing

ISBN-13: 978-3-322-83058-6   e-ISBN-13: 978-3-322-83057-9
DOI: 10.1007/978-3-322-83057-9

Preface

The present volume of articles grew out of a workshop on language engineering, or better: the principles of an engineering approach for the development of language products, that was initiated and organized by Gerhard Heyer in May 1991 on the premises of the Reimers Stiftung, Bad Homburg. This is the list of persons that participated: Dr. J. Arz (Darmstadt), Dr. H.-U. Block (München), Dr. P. Bosch (Heidelberg), Prof. Dr. U.L. Figge (Bochum), Dr. Ch. Galinski (Wien), T. Gerhard (Luxembourg), Prof. Dr. W. von Hahn (Hamburg), Dr. H. Haugeneder (München), Prof. Dr. R. Hauser (Erlangen-Nürnberg), Prof. Dr. G. Heyer (Leipzig), Prof. Dr. H. Kamp (Stuttgart), Prof. Dr. Krause (Bonn), Dr. J. Laubsch (Palo Alto, USA), Dr. D. Rösner (Ulm), Prof. Dr. H. Schnelle (Bochum), Prof. Dr. J. Siekmann (Kaiserslautern), Dr. O. Stock (Trento, Italy), Dr. G. Thurmair (München), Dr. H. Trost (Wien), Prof. Dr. H. Uszkoreit (Saarbrücken), Prof. Dr. W. Wahlster (Saarbrücken), Dr. M. Zoeppritz (Heidelberg).

All contributions were of very high quality, and the discussion has certainly been very stimulating and fruitful at a stage where the very notion of language engineering has not yet become a (rather polysemous) commonplace. Walther von Hahn and Rainald Klockenbusch were so kind as to immediately commission the proceedings, and one of the very first books on language engineering could soon have appeared. In the months to follow, however, editorial work was delayed for at least two reasons. Not only had there been contributors to the workshop who did not in the end contribute to this volume, but, more importantly, there were also changes in the professional environment for both the editors that forced them to existentially concentrate on retaining natural language processing as their professional orientation. It is one thing to discuss how a successful language engineering should look like, but it is quite another thing to really build commercially successful language products. Time to reflect on the general principles of our discipline became a privilege that in a true sense of the word had to be earned first.

In view of the many changes of European computer and software industries during those four years, and considering the obvious need for most of us who work on natural language processing to turn it into a profitable discipline, the notion of language engineering may have even more practical relevance today than it did in 1991. When re-reading the present collection of papers I became convinced that they would still make a valuable contribution to the present discussion, and herewith take the risk of publishing them in a state at which the book had already been in 1993.

Many thanks to Walther von Hahn and Rainald Klockenbusch for their patience and support, many thanks also to my secretary at Leipzig, Mrs. Renate Schildt, without whom the final attempt to get this book published would not have succeeded.

Leipzig, March 1995 Gerhard Heyer

Contents

Perspectives

Introduction

Hans Haugeneder
Siemens AG
Corporate Research and Development
Otto-Hahn-Ring 6, D-81730 München
Hans.Haugeneder@zfe.siemens.de

Scope

Natural Language Processing has been a scientific enterprise for 40 years. In the last decade and half, however, the interest in Applied Natural Language Processing has increased steadily. The reason for this development can be seen in the advances in various areas of computational linguistics on the one hand and both the availability of inexpensive and powerful computing machinery and the growing demand for various types of natural language functionalities as part of software systems on the other hand. This increase in practical and industrial interest has also been accompanied by substantial funding. And although the natural language community has been more careful (perhaps due to the experience of the ALPAC report in the sixties) than researchers in other fields of Artificial Intelligence — where the creation of unrealistic expectations led to a substantial scepticism followed by a decrease of support — fairly high expectations with respect of the near and mid-term exploitability of natural language products do exist.

The contributions in this collection can be viewed as an attempt to provide a realistic view of today's state of the art in Natural Language Processing, taking into account the experience of both academic and industrial researchers. The major issues that are tackled throughout this volume can globally be related to the following issues:

- What is needed to turn the process of creating language products into an engineering discipline?

- What are the specific methodological prerequisites for such a process?

- What are the perspectives and what is the strategy for the development of a language technology?

With these aims in mind the contributions in this collection are not intended to present specific natural language systems as such. They rather provide a more reflective view concerning different aspects in the design and use of natural language components and systems. Nevertheless, a quite representative selection of the application fields for natural language technology is addressed, such as speech understanding systems, natural language interfaces to date bases, natural language command and help systems, grammar and style checking in the context of machine translation and multi-modal interactive information systems.

Most of the contributions emergered out of a workshop entitled "Language Technology — Natural Language Processing in Application"[1], held in May 1991. The workshop was supported by the Werner-Reimers-Stiftung, which provided the stimulating atmosphere of their Bad Homburg facilities.[2]

Engineering Aspects

In the first contribution G. Heyer sketches the elements of an engineering approach to building natural language processing components based on the assumption of the need for a different theoretical foundation for language engineering than the one given in theoretical and computational linguistics. The provision of a holistic notion of efficiency which also takes into account non-linguistic factors is considered to be a main desideratum of such a genuine foundation. This, however, implies a different orientation than the one that is prevalent in theoretical computational linguistics with its primary goal of high linguistic generality, since from an engineer's needs a set of tools for the creation of a specific solution for a certain type of language product is the real issue. In order to support such an engineering approach a theoretical framework for optimizing between linguistic and non-linguistic requirements needs to be developed. Steps towards such a framework are discussed in the context of applications of both the interface and autonomous task type. The provision of reusable linguistic knowledge sources which can be compiled into specific applications is envisaged as one core element of such an approach. The details

---

[1] The original German title was "Sprachtechnologie und Praxis der maschinellen Sprachverarbeitung"

[2] We would like to thank the foundation for the hospitality and specifically H. v. Kroszig for his steady support.

are exemplified in the field of computational lexicology, where as a first step the provision of linguistic data embedded in active electronic media is identified. Ergonomic considerations as a particularly critical factor for engineering interface products are explored in the context of an evaluation of the natural language help system GOETHE, which supplies help for the Unix file system. They indicate the limited importance of linguistic phenomena for the quality of the overall language product. Thus, in order to be able to successfully create natural language products, one has to integrate a number of quite different, even competing aspects like software engineering, software ergonomy, linguistic theory and the specifics of human-computer interaction (see also Krause and Figge in this volume for contributions to this issues) into a comprehensive language engineering technology.

M. Zoeppritz discusses natural language systems from the perspective of software ergonomics with natural language database query as the underlying scenario. After a compact survey and a synthesis of relevant user studies the importance of those ergonomically relevant features that are transparent to the user are discussed: habitability of the system's coverage, the integration of the natural language functionality in the application environment, the handling of errors having their origin in the natural language component and the availability documentation. Finally the importance of ergonomic features on the side of the process of developing natural language components is stressed and it is argued for development tools with an expressively high specification level for the linguistic knowledge bases, powerful and reliable debugging facilities and a systematic evaluation methodology.

J.A. Banwart and S. I. Berrettoni propose the adoption of methods that have been successfully used in the software development process for the creation of natural language products. As a highly critical and vital phase in such a development process they identify the test phase. As particularly important aspects the use of different test strategies, the interpretation of test results and the requirements for test tools are identified and described in detail. Although the testing procedures taken from standard software development practice can be considered as a useful starting point, they do not cover all aspects of the development of language products (see also Gerhardt in this volume). This is shown with respect to the process of testing multilingual natural language systems when compared to multilingual non-language software products.

Methodology

J. Krause approaches the question of how to integrate natural language and graphical interaction facilities into an adequate model of human-computer interaction. One central problem that has to be clarified with respect to this issue is the characterisation of the nature of natural language that actually occurs in the interaction between man and machine. Contrary to the hypothesis of the identity of the user's communicative behaviour in scenarios of man-machine interaction with that in human-human interaction the existence of a specific language register, "computertalk", is argued for on the basis of the empirical results gained in the DICOS project. As further support of this thesis the results of studies performed in the last fifteen years are taken into account. Some of the examples of computertalk observed in the experiments are not confined to - more or less - minor differences from the standard use of human language, they rather exhibit fundamentally divergent use of of language. This is explained by the human speaker's (tacit) assumption that the computer as recipient of his utterance is not attributed the faculty of understanding human language.

In their contribution H.-U. Block and S. Schachtl give an illustrative example of a development strategy for linguistic knowledge sources which takes into account the linguistic data given in a target application, thus reflecting the actual needs with respect to the coverage to be achieved. In particular they address the problem of providing an adequate grammatical description of the use of spoken language in a dialogue scenario, not with the pretension to provide a theory of spoken language, but rather aiming towards a characterization of the syntactic properties that are specific for such a dialogue application. Based on the analysis of a quite broad sample - the empirical basis was gained by taking into account the results of serveral independent data collection efforts - a linguistic classification of spoken language phenomena is provided, comprising those grammatical construction types that are relevant "in the imminent task of understanding spontaneous speech dialogues". The results of this analysis document the substantial efforts that are necessary in order to transfer the linguistic knowledge for understanding written text to the area of understanding spoken language. In more general terms this work also exemplifies the importance of a data driven engineering approach for delivering descriptive linguistic knowledge sources of practical relevance as opposed to a

strategy of defining the coverage of natural language dialogue systems on the basis of unwarranted projections based on abstract grammatical considerations.

In his reflections on the human language and the computer U. Figge distinguishes three different types of relation between human language and the computer: the computer as tool for the study of human language, the computer as a model of human language behaviour and finally, the computer as a technical device which is controlled by means of human language. Whereas in its first role the computer is not unique - it is just the latest and most powerful tool-, the two latter views are considered to constitute genuine new scientific and technological challenges. Their potential with respect to applicability is explored in more detail. Concerning the use of natural language as a command language it is argued that human language as a particular instance of a semiotic system can only be used fruitfully if it is integrated tightly with specifically designed semiotic systems for man-machine communication. This gives rise to "computational semiotics" as a new, still to be established subdiscipline of semiotics.

Application

The fundamental question of how to design the architecture for natural language systems is discussed by W. v. Hahn and C. Pyka with special focus on speech understanding systems. The basic hypothesis underlying their approach is that an integrated processing model for language and speech must be incremental, synchronous and (almost) deterministic. After a clarification of the notion of "architecture" and a discussion of the major architectural models used in natural language systems, the conceptual layer of an architecture which is based on an interactive incremental model of computation is presented. As a core feature of the architecture message passing is the only modality for the interaction between the single components dedicated to specific levels of analysis. Within this modality of message passing two fundamental types of communication are distinguished. On the one hand hypotheses (i.e. local results) created by a specific component can be made available by transmitting them from the originating component to another one. On the other hand data can be requested from a component in order to resolve local ambiguities by non-local information. Finally, the architecture's benefits with respect to incrementality, determinism and synchronicity in the process of language analysis are shown.

J. Laubsch presents an approach for creating application specific interfaces from a domain independent semantic representation, which is particularly relevant for the development of natural language front ends for different domains and application interfaces. The domain independent representation formalism used, NLL, is a logic-based formalism in the spirit of the language of generalized quantifiers. It encodes the meaning of natural language expressions in a rigid, well-defined fashion. The process of mapping representations expressed in this formalism into the one that is "understood" by the application is performed through various types of transformations, like simplification, disambiguation, and domain specific inferences. As an example of programming an application interface on the basis of the approach proposed the mapping of NLL to the relational data base language SQL is described.

Although natural language interfaces to data bases have always been viewed as one of the most promising candidates for practical and productive application of natural language technology, their acceptance has been quite modest up to now. Based on the experience with Datenbank-DIALOG, a German language interface to relational databases, H. Trost presents a comprehensive overview of the factors relevant for the success of failure in real world applications and discusses their role in creating easy to use natural language data base access facilities for different domains. From the viewpoint of a system's engineer - as the producer of a specific installation - a high degree of portability with respect to a number of different levels is of central importance (see also Laubsch in this volume). From the user's point of view the linguistic coverage, habitability and robustness are the critical features. The first of these, coverage, determines the overall acceptance not as strongly as generally assumed. Despite the lack of general approaches to handle very commonly used expressive means like elliptic utterances, discourse particles and anaphora sufficient coverage can be provided. The second dimension, habitability, i.e. the specification of the sublanguage used, constitutes a core feature and at the same time a challenge, since defining the restrictions which a human user finds natural is not yet understood in a fully systematic way. The third dimension, robustness, has to comprise both the creation of sensible system reactions and retracting errors. From an engineering point of view it is the well-balanced combination of the various features that leads to stable systems with predictable behaviour enabling a successful transfer of natural language interfaces to the end user.

The design of a component for grammar and style checking to support good technical writing, the development of which is performed in the context of a translators workbench, is the topic of G. Thurmair's contribution. After giving the rationale for the use of controlled grammars - they are considered as implementation of guidelines for document production - the author discusses the fundamental dimensions of stylistic guidance, comprising layout control, text presentation conventions, the use of terminology and the preference of grammatical features, with the latter aspect being focused within the approach. The system's operation is based on a model that compares analysed linguistic entities with predefined patterns of deviance. An input sentence analyser provides a full phrase structural description of the unit under consideration, a matcher compares these with a set of structures that are considered deviant (the illformed structures repository) and an output generator creates the diagnostic information. Thus, the approach chosen implies the use of quite substantial linguistic sophistication concerning the linguistic knowledge sources and computational mechanisms. After a presentation of the practical results gained in several experiments a strategy for embedding such a component into a productive environment such as the METAL machine translation system is sketched with respect to both architectural and ergonomic issues.

In their contribution J. Arz et. al. discuss the essentials of a strategy needed for marketing natural language products. One of the most important factors is the provision of a smooth and functionally sound integration of natural language facilities in the overall product, a feature that has been carefully taken care of in DOS-MAN, a natural language help and interface system for DOS. Only if this issue is taken seriously the failures that occurred in launching stand-alone natural language products can be avoided. Finally, a general insight into the process of commercialisation of natural language products is given by means of the presentation of the ecomomic history, the product spectrum and the client structure of TRANSMODUL, one of the few companies mainly dedicated to natural language products.

Perspectives

The major part of the following contributions have their origin in and have been partially stimulated by a panel discussion during the "Language Technology" workshop[3]. The topic of this panel "Language Technology - Myth or Reality?" was pretty broad and so is the scope of the contributions, which are extended versions of the panel statements made by the authors. Some of these contributions focus on specific aspects while others aim towards a comprehensive treatment of the issues raised within this panel discussion. In order to put the contributions into the overall context the reader is referred to the original questionnaire which is presented below.

1. State of the Art in Language Technology

    (a) Does the state of the art in natural language processing deserve the predicate "technology"? (i.e. Do we already have language "technology"?)

    (b) What are the strengths and the shortcomings of today's language technology along the following dimensions:

    - linguistic knowledge specification (formalisms, knowledge sources)

    - procedural models (bidirectionality, efficiency)

    - development tools

    (c) What has been the rate of progress in language technology over the last ten years? Which are the fields of major progress?

    (d) What are the application areas in which today's state of the art in language technology can produce useful and competitive products?

2. Generic vs. Application-specific Orientation

    (a) Is it useful and feasible to strive for a generic application-independent language technology?

    (b) Are there application areas for which well-defined subsets of language processing facilities are sufficient?

3. Design and Evaluation

    (a) Do we have (need) a stable design methodology for language products?

    (b) Do we have (need) a stable evaluation methodology for language products?

---

[3] The participants at this discussion were: H.-U. Block (Siemens München), P. Bosch (IBM Stuttgart), Ch. Galinski (Infoterm Wien), T. Gerhardt (CRETA Luxembourg), D. Rösner (FAW Ulm), O. Stock (IRST Trento)

(c) What can we learn from other areas (e.g. software engineering) in design and evaluation?

4. Future Developments in Language Technology

(a) What are the commercially driving application areas for future developments in language technology?

(b) What are the research topics that are expected to have the most impact for a language technology?

(c) Is it productive or dangerous to attack highly complex application areas (as combining machine translation and speech understanding?)

In his contribution P. Bosch introduces a broader scope to the notion of language technology. He argues for a language-and-information technology based on the assumption that any computational model of natural language has to take into account its deep embeddedness in commonsense knowledge, which by its nature is non-linguistic. As a consequence of this view "language" is understood in a quite comprehensive way that exceeds the perspective of linguistics centered around linguistic form. It is considered "a semantically interpreted medium for representing and processing information". As the most promising application area for a language technology of this type natural language interfaces are seen. They constitute a classical example for the need of integrating linguistic and world knowledge. Thus it is argued that natural language systems are instances of knowledge based systems, leading to the conclusion that language technology is only thinkable as a part of a knowledge technology.

The key issue in O. Stock's contribution is the question of naturalness of natural language based human-computer interaction, which is often assumed either implicitly or on the basis of a quite naive mixture of cognitive and ergonomic arguments. Rather than arguing for natural language as the only and most advantageous communication tool he points out that both multi-modality and hypermediality complement natural language in a sense that is very relevant in practice, since their combination can increase the habitability in intelligent interfaces. The benefits of such an approach of integrating different media are discussed in the context of AlFresco, a multi-modal interactive information system for frescoes.

On the basis of a characterization of the notion of technology as the "description of procedures and methods in order to form a product" T. Gerhardt defines language technology with reference to the abstractness of the products obtained, as compared to more physically embedded products such as hardware. Under this view the products of language technology are software products enhanced with different types of linguistic functionality. After an explication of the relation between the design and evaluation phase in the overall development process and the presentation of a model of evaluation based on the similarity between them, it is argued that the parallelism between these two phases is misleading if taken as global methodological framework for evaluation, since it does not take into account the product's end users. Despite their classification as software products with specific linguistic augmentations it is an insufficient approach to evaluate natural language software in the same way as standard software products. This leads to a number of difficulties which are exemplified with reference to spelling correction and machine translation products.

In his contribution on the state and perspectives of language technology D. Rösner identifies the increase of the coverage and improved reusability of linguistic knowledge sources such as lexica and grammars as a central issue for a successful transition from academic computational linguistics towards a language technology. The growing awareness of the vital role of reuseable knowledge sources and the fact that this area is being tackled within several dedicated initiatives, both on national (USA, Japan) and European level is seen as an indication of growing maturity of the field. Concerning the future application potential for natural language products dealing with multilingual aspects are viewed as the most promising option, since they are driven by real needs. These have their origin in the growth of global, multi-national markets where systems providing various types of machine translation functionality (for product documentation etc.) will constitute a competitive factor on the cost and on the user side.

Starting from the observation that research in pure linguistic is only partially relevant for the development of language industries - which has the consequence that the provision of scientific solutions is not sufficient at all for generating language products - H. Schnelle puts the current efforts towards a language industry into the context of the historico-cultural development concerning the methods and tools for encoding and reproducing language. The present stage in this development process is characterised by the storage of language products on active media which can be manipulated by computers with enormous flexibility. As one rather powerful type of manipulation methods for the computation of linguistic form have been provided by computational linguistics. In this subfield of computational linguistics there has been substantial progress, as documented by the state of the art in parsing and morphological analysis for example, which has led to a stage where the development of technological products based on computational models of linguistic form is feasible. Instead of waiting a few decades for comparable progress in the still underdeveloped disciplines of semantics and pragmatics, an evolutionary development strategy for a language industry is argued for, aiming towards the exploitation of restricted computational models of linguistic competence. Such a strategy facilitates the delivery of stepwise improvements for existing widespread language products, such as dictionaries and grammars, where radical enlargements of their functionality can be achieved. To this end the evolutionary approach has to be given more attention to within computational linguistics. It should be persued actively and in parallel to the currently favoured revolutionary strategy which has taken the challenge of developing a full computational model of human language competence. This can, however, only be considered a long term task.

Synthesis

At the current stage language engineering cannot be considered a settled member in the realm of the well-established engineering disciplines. This is a consequence of the fact that a language technology in the broad sense which subsumes the computational perspective of providing methods to construct components and systems that process natural language(s) both spoken and written with different granularity for different purposes has not yet been fully developed. This is not surprising at all if one takes into consideration the complexity of the domain and the youth of the technological field and its contributing sciences such as theoretical and computational linguistics, computer science and artificial intelligence, ergonomics and software engineering, as compared to the classical engineering disciplines. On the other hand a language engineering discipline starts to emerge with technology push and market pull as the two driving forces. This is not to say that all the technology needed is already there nor that the information technology market is crying for nothing but natural language products; there are, however, substantial technological potential and real practical needs. The contributions in this volume address many facets concerning the current state of the art, the problems of practical exploitation and the strategy for choosing the right route to stengthen the technological dimension of natural language processing. And although quite a number of controversial and critical issues have been raised, there are some topics and positions which have been addressed throughout the various contributions and which are summarized below in the following six theses:

- Although computational linguistics constitutes a core element for language products by providing underlying natural language processing functionality, a view confined to it is too narrow for an engineering approach for the development of natural language products. The quality of natural language products is not identical with their linguistic functionality, nor is their acceptance exclusively determined by it.

- Research in computational linguistics should not confine itself to the goal of providing a full computational model of the human language faculty, despite the fact that this is undoubtedly an important long term orientation both scientifically and technologically. The development of (even highly) specialized linguistic functionalities is a much more promising mid-term strategy for developing useful language products.

- Human-computer interaction by means of natural language is not isomorphic to human-human communication. The differences have to be described systematically and taken into account in the design of systems for human-computer interaction by means of natural language.

- Natural language is only one part of the human semiotic system, though important and rich in its expressive means. The integration of natural language with other modalities is a core challenge, both from the theoretical and the practical perspective.

- A methodology for language engineering cannot be established by purely adopting methodology of software engineering, although many of of the techniques used there can be applied fruitfully.

- Evaluation of natural language products has to be taken into account much more intensively. A specific comprehensive methodology for systematic evaluation for systems with natural language functionality still has to be provided.

If taken into account seriously in the next decade of research and development activities in Applied Natural Language Processing it is our conviction that these orientations will help to shift into a higher gear in language engineering.

Elements of a Natural Language Processing Technology

Gerhard Heyer
Universität Leipzig und Electronic Publishing GmbH Nürnberg

Contakt:
Universität Leipzig, Institut für Informatik
Abteilung Automatische Sprachverarbeitung
Augustusplatz 10/11, D- 04109 Leipzig
heyer@informatik.uni-leipzig.de

1   Introduction

In order to meet the growing expectations raised by serious market studies for the remainder of the decade, it is commonly agreed that natural language processing must be based on a sound theoretical basis that allows for a simplified construction, adaptation, or modification of natural language programmes to particular tasks. The task is quite a general one, since to the extent that natural language products are software products, the development of natural language programmes needs to take into account at least the dimensions of software-engineering, ergonomics, and linguistic functionality. Successful commercialization of natural language processing is not a natural consequence of more sophisticated linguistics, but primarily requires an efficient management of the linguistic software development cycle, and a clear conception of the use of natural language as a means of communication in general, and human computer interaction in particular.

In what follows I shall propose, exemplify, and discuss what can be called the *engineering approach* to the above problem of a methodology for the construction of natural language processing programmes. In essence, the idea of language technology as based on the engineering approach, taking as starting point a standard definition of engineering science in general (Pahl/Beitz 1986:1), conceives of language technology as a genuine part of computational linguistics that uses results from theoretical linguistics and computer science in order to find solutions to technical problems in all areas of natural language processing, and implements them in an optimal way given particular linguistic, computational, and financial constraints.

As we shall see, the present day cognitive paradigm in computational linguistics is only of limited use to this engineering approach. The problem, in my view, is *not* that of a transfer from science to technology, but the fact that *different theoretical foundations* are needed for applied computational linguistics than are provided by a computational linguistics oriented towards cognitive linguistics, despite some overlappings in certain areas. The two approaches to natural language processing differ both in methodology and objectives: While cognitive linguistics strives for general solutions based on well established theoretical paradigms with the aim of extending a paradigm to other classes of linguistic phenomena, language technology is problem driven, and tries to find optimally engineered solutions to problems arising in specific application tasks. In general, language technology aims at an optimization of theoretical costs and practical benefits, while any such optimization is alien to a purely theory driven concern with language. In practice, the main task of language technology is to manage efficiently linguistic complexity in terms of breadth of data and depth of processing required, while from a point of view interested in linguistic competence only, one can rest content when such complexities have been sufficiently described.

Although there appears to be a growing awareness in the scientific community about the fact that small-scale laboratory prototypes cannot simply be turned into language products without additional engineering efforts, there does not appear to be a clear understanding yet of the principle factors that are constitutive of language technology. The paper is intended to contribute to this discussion about the applicability of a cognitively oriented computational linguistics, and to present a first sketch of elements of a natural language processing technology.

2   Software Engineering for language products

User and producer of language products conceive of language products primarily as software products. That means, language products need first to satisfy software quality requirements like *user habitability* (e.g. coverage, robustness, reliability, speed) and *implementor productivity* (e.g. portability to different software and hardware environments, changeability to other natural languages, modifiability of grammar, testability, and verifiability) (cf. Carbonell 1986:162).

In practice, user habitability requirements need to be relativized to particular hardware and software environments that in most cases imply performance restrictions. Thus, contrary to the notion of efficiency as commonly employed in complexity theory and related discussions of parsers, we employ a holistic notion of *efficiency* of a natural language processing programme, and take it to mean the way how the programme performs its task by exploiting all available resources, including, in particular, hardware constraints (RAM, ROM, CPU), software environment (operating system, availability of efficient compilers), and detail and coverage of lingware (lexicon, grammar, parser, knowledge structure of application domain). As always in real life, one-sided optimizations will lead to sub-optimal behaviour of the whole system. Holistic efficiency of a language product, therefore, always is the result of optimizing between the available linguistic and non-linguistic resources relative to the functional specification of a language product.

The point also holds for the efficiency of the linguistic processing modules themselves, neglecting other hardware and software constraints for the moment. Assuming an architecture where at well defined points during the natural language processing first a syntactic structure, then a semantic structure, and finally a knowledge structure (depending on the application) is created, optimization on the complexity of each structure will not only have draw-backs on its computability (cf. Habel 1988:208), but will also jeopardize the efficiency of the whole system, as structures from one module may be passed on to the next without any provision of adequate further processing. Thus, of all those sentences that can syntactically be parsed in depth, only a limited number can be assigned an equally complex semantics, given the present state of computational semantics, and for all those sentences for which we have, say, an intensional second order predicate logic semantics, we will then have again only a limited number of sentences for which we can provide an adequate knowledge processing. In effect, we leave information generated by other modules wholly, or partially, unused, as it is too complex to be further processed. At any rate, the whole system's behaviour will be less than optimal: Either it will be able to cope in depth with only a very limited set of sentences (determined, in fact, by the knowledge processing capacities), or it will be able to process its input data in breadth, but in doing so produces a substantial amount of redundant information (Heyer 1990:39).

The notion of efficiency also is central to the development philosophy of natural language processing software. While a holistic design reduces interface problems and increases efficiency, it generally leads to software that is not easily ported, both to other platforms and to other natural languages or applications. In order to increase implementor productivity, the nowadays preferred methodology in computational linguistics is a modular design of natural language processing programmes. (The preferred modular approach also supports nicely the view that a computational linguist only needs to know how to do linguistics on a computer, but does not need to know anything about the way how his formalizations can be efficiently implemented).

What the modular design gains in portability, however, it looses in efficiency, because it assumes, in its extreme form, a level of general and all purpose natural language processing software that is to be used in all kinds of applications. But any such general linguistic processor, quite like Newell and Simon's general problem solver, will always process the data in a specific application much less efficiently than a system tailored to the specific application and based on a holistic design. From an engineering point of view, therefore, neither approach is optimal.

In order to provide a cost efficient basis for all kinds of natural language processing programmes, and as input for tools by which holistic natural language processing solutions can be compiled, I rather imagine an approach that results in multi-functional, reusable linguistic software on all levels of linguistic knowledge, viz. lexica, grammars, and meaning definitions. Let us call this methodology for the construction of natural language processing programmes the *compilation approach* (see figure 1).

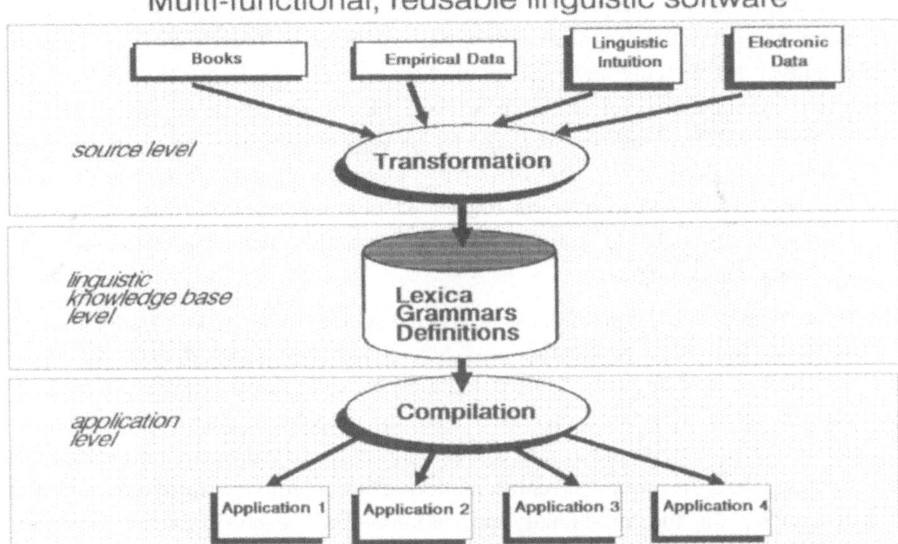

Figure 1

The intuitive idea of the compilation approach is to construct highly efficient and holistically designed natural language applications on the basis of linguistic knowledge bases that contain basic and uncontroversial linguistic data on dictionary entries, grammar rules, and meaning definitions independently from specific applications, data structures, formalizations, and theories.

To the extent that linguistics is a more than two thousand years old science, there is ample theoretical and empirical material of the required kind available in form of written texts and studies on the source level of linguistic knowledge, or can be (and is in fact) produced by competent linguists. However, very little of this knowledge is also already available on electronic media. Thus, the very first task of language technology, as Helmut Schnelle has recently put it (Schnelle, this volume), must be the transformation of available linguistic data from passive media into active electronic media, here called lexica, grammars, and definitions on the linguistic knowledge base level. In terms of implementation, such media will mainly be relational, object-oriented, or hypermedia data bases capable of managing very large amounts of data. Clearly, in order to be successful, any such transformation also requires formalisms to be used on the side of the linguists for adequately encoding linguistic data, the linguistic structures assigned to them, and the theories employed for deriving these structures. Moreover, within linguistics, we need to arrive at standards for each such level of formalization. In actual detail, therefore, the first task of language technology is quite a substantial one that can only succeed, if the goal of making linguistic knowledge technologically available is allowed to have an impact on ongoing research in linguistics by focussing research on formalisms and standards that can efficiently be processed on a computer.

Now, to complete the picture, assuming that such a linguistic knowledge base is available, individual applications are to be constructed, adapted, or modified on the basis of this linguistic knowledge base by selectively extracting only that kind of information that is needed for building the specific application, and by compiling and integrating it into the application specific data structures. Details, coverage, and the compiled representation of the linguistic information depend, of course, on the individual applications. The second task of language technology, then, consists in providing a general methodology for such a selection of the required linguistic knowledge, and the definition of its optimal data structure representation.

To put this picture to work, let us briefly look at how it can be applied to issues in the area of computational lexicography as presently discussed in ESPRIT II project MULTILEX (see figure 2) (Heyer/Khatchadourian/Modiano/Waldhör 1994). Whether or not the compilation approach can also be extended to grammars and meaning definitions as the main other two areas of linguistic knowledge remains an open question at present.

On the source level, there are printed dictionaries, text corpora, linguistic intuitions, and some few lexical data bases. In order to make these sources available for language products, we first need to transform the available lexical data according to an exchange standard into a representation standard on the level of a lexical data base (for each European language). The exchange standard proposed by MULTILEX is SGML, following recommendations from the ET-7 study on reusable lexical resources (Heid 1991). The representation standards proposed for the different lexical levels are the Computer Phonetic Alphabet (CPA) for the phonetic level, the ISO orthographic standard for the orthographic level, and a typed feature logic for the morphological, syntactic, and semantic level. On this functional view, implementation details of the data base are irrelevant as long as it allows for an SGML communication. The

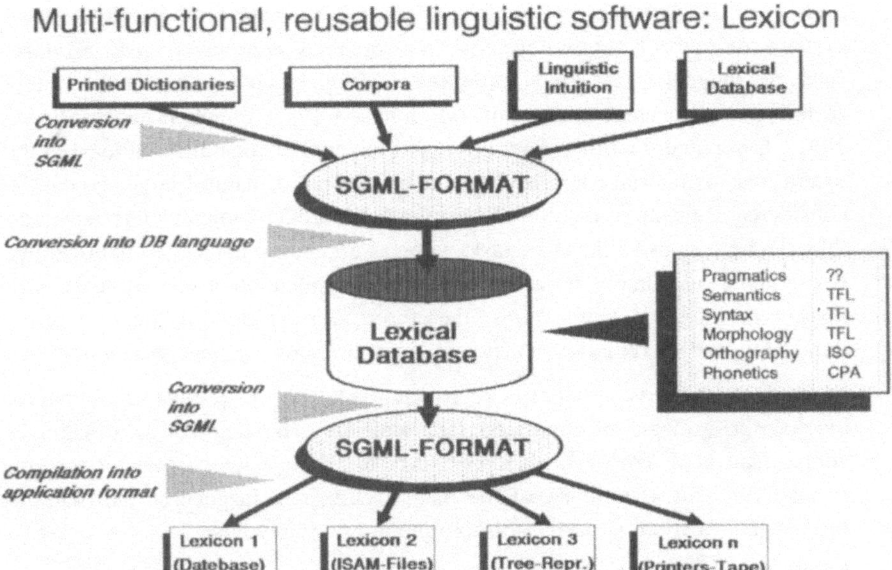

Figure 2

master lexical data base is then being used to develop specific lexicon based language products such as multilingual electronic dictionary applications for human and machine users in the area of automatic and semi-automatic translation support, highly compressed multilingual spelling correctors and language checkers, highly compressed lexica for optimizing handwriting recognition, etc. Needless to say that each of these products uses different kinds of linguistic information as provided by the lexical data base, and also uses different ways of representing the linguistic knowledge most efficiently with respect to particular hardware and software constraints. In view of the distinction drawn below between *interface* and *autonomous tasks products*, however, it is to be expected that the compilation approach is at present best suited for autonomous tasks language products only.

3   Ergonomics of language products

Natural language products are commonly distinguished into *interface products* such as main frame or PC dialogue systems for data base queries, expert systems, or decision support systems, and so-called *autonomous tasks products* such as language checking tools for intelligent text processing, content scanning, and machine translation (cf. Johnson 1987, Engelien and McBryde 1991). From an ergonomic point of view, both applications differ at least to the extent that for interface applications the processing of natural language can be considered a means to the end of an improved human computer interface, and thus can be seen as an interface task, whereas for autonomous tasks applications it can more adequately be considered as an application itself. At least with respect to interface applications, therefore, consideration of the ergonomic dimension is also crucial to an efficient engineering of language products.

In practice, there is a widespread assumption that the provision of natural language as a means of communication with the computer will as a matter of course lead to a more task adequate, natural and efficient user interface (cf. Krause 1982 for a summary of the main arguments). In general, and without further qualification, however, this assumption is simply false.

Looking in detail at the problem of ergonomic design, it first needs to be noted that in every particular case, natural language user interfaces compete with command languages, menus, and graphical user interfaces (see also Shneiderman 1987:167, Obermeier 1989:191). Each user interface is implemented in an environment comprising at least a periphal, a terminal, and a CPU. In terms of the minimal hardware requirements for the different user interfaces, natural language interfaces compare to commands in that they only require a keyboard and an ASCII terminal (for the purpose of the natural language interface), but in terms of computing power require about as powerful a machine as is needed for running a graphical user interface such as WINDOWS. In other words, in the evolution of user interfaces, natural language interfaces represent the unusual case that they require a comparatively cheap peripheral and terminal, while they simultanously require a comparatively expensive CPU.

The main task of ergonomic design for language products consists now in determining which features of an application require from the language technology point of view which features of a natural language interface, given certain hardware and software constraints. This question can at present in most cases only be decided on an empirical basis.

As an example, let us consider a context dependent natural language help system for UNIX developed in 1990 by TA Triumph-Adler for Olivetti Systems and Networks in the framework of ESPRIT I project 311, ADKMS. Comparable to other UNIX help systems like UNIX Consultant or SINIX Consultant, the GOETHE system offers the user a context dependent mode of natural language command and help request interaction with the UNIX file system based on a static and dynamic knowledge base about UNIX, and, in a given situation, actual user-UNIX interactions. In its present state of implementation, the system comprises 70 UNIX programmes, an integrated parser with a dictionary of about 800 entries, and adequately recognizes about 60% of natural language user input within an average response time of 10 seconds on an Olivetti LSX machine (Heyer/Kese/Oemig/Dudda 1990). In order to evaluate the theoretical costs for the linguistic modules and their practical ergonomic benefit, an empirical evaluation was carried out on 28 subjects, covering significant portions of novice, casual, and expert UNIX users, that were asked to perform 39 well defined tasks using plain UNIX, the natural language interface GOETHE, and a mock-up graphical user interface (Möller 1991). The main results are as follows:

(1) General ergonomic deficiencies of typed natural language input. With respect to GOETHE, users most frequently report problems concerning indadequacies like slowness, typing efforts, and complicatedness (see figure 3).

We interpret this result as a general complaint about ergonomic deficiencies of interfaces based on natural language input typed in on a keyboard that hold irrespectively of the quality of the linguistic processing modules. Some of these problems may be overcome with a general real time processing of speech input. However, instead of investing more into the natural language processing, in many cases it may be from an ergonomic point of view more effective to provide the natural language interface with smarter editors, correction functions, lexicon browsers, and the like.

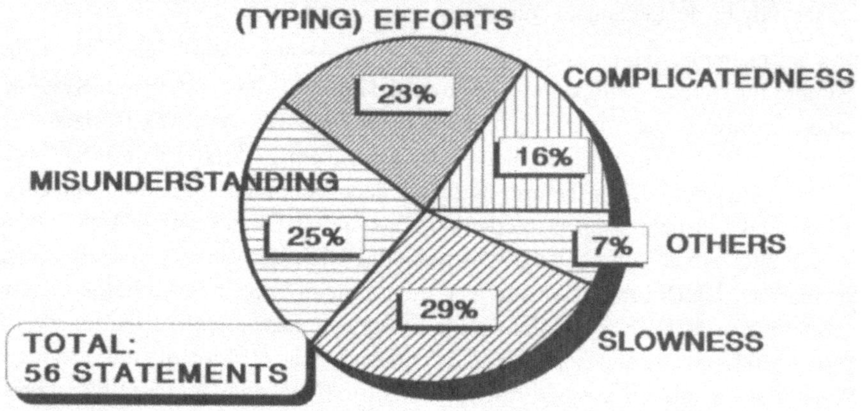

Figure 3

(2) Semantic profile of user interfaces. Considering the weighted scores of user assessment concerning UNIX, GOETHE and the graphical user interface in the domain of application with respect to the standard ergonomic parameters *explainability, controllability, task suitability, predictability, fault tolerance,* and 15 other related criteria, the natural language interface turns out to be always weaker than the graphical user interface, and better than the plain UNIX command language only with respect to *explainability, controllability, lucidity, ease of learning, vividness, transparency,* and *simplicity* (see figure 4).

The weakness of the natural language interface apparently has to do with the fact that natural language is a general purpose means of communication and optimized neither for efficiency, nor for particular tasks in the area of human computer interaction, although some negative performance assessments, like *fault tolerance, speed,* and *saving of time,* may have to do with the particular implementation. At any rate, the weakness can also be considered a strength in that natural language is easily applicable in many situations, and, in particular, in situations where more task optimized means of communication have broken down. Considering the estimated suitability of user interfaces for particular tasks, this interpretation is also confirmed by the fact that context help and task help are the only tasks where GOETHE has been judged

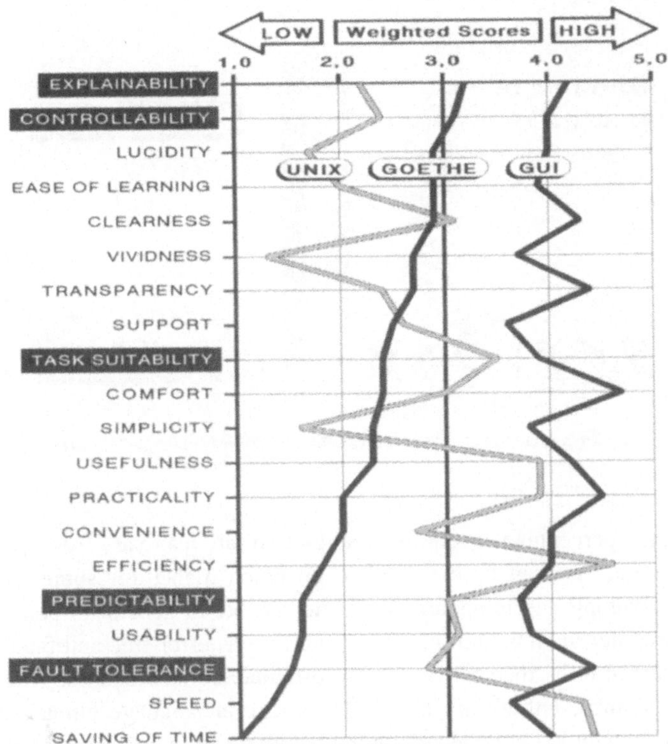

Figure 4

superior to UNIX and the graphical user interface. From the point of view of language technology, it may be important, therefore, to distinguish between a command entry and help request mode for natural language interface applications, and to consider, where possible, a combination of GUI and natural language interface components where natural language is mainly used for formulating general help requests.

(3) Required recognition rate for user input. While in computational linguistics research oriented towards cognitive linguistics, there is a tendency to always expect grammars and parsers to yield a 100% recognition rate, user

reaction in the course of the GOETHE evaluation surprisingly indicates that the average percentage of required recognition rate may be substantially lower, and in the specific application is in fact 91% (see figure 5).

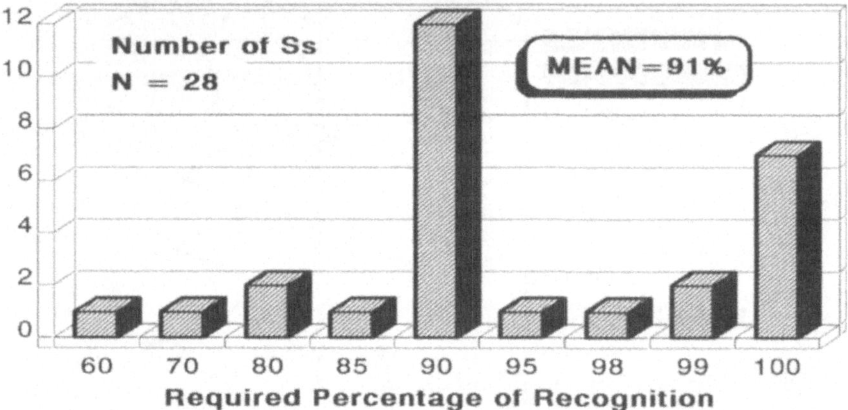

Figure 5

Although the percentage of required recognition rate may vary from application to application, and can be expected to be much higher for some translation tasks, for example, users apparently do not expect in general to communicate with a computer via a natural language interface as error-free and fault tolerant as is presumed to be the case in human-to-human communication. The finding can be explained, I think, by the fact that natural language human-computer interaction cannot be modelled along the paradigm of linguistic communication among humans, as is often assumed (e.g. Kanngießer 1989), but must be understood as a communication *sui generis*. As Krause has recently shown on the basis of substantial empirical investigations (Krause 1991, cf. also this volume), human-computer communication can be said to differ from ordinary human-human communication at least in terms of the envoked sublanguages and language registers. But if *computer-talk*, a language register comparable to baby-talk or foreigner-talk, leads to simplified and systematically distorted user input, it is plausible to assume that it also leads to a higher tolerance on the user's side towards incomplete or erroneous processing of language data on the side of the computer. From the language technology point of view, this finding will have the important consequence that if users do not require a 100%

recognition rate of natural language input for a particular application, accuracy of the natural language recognition may be traded against broader linguistic coverage, less memory consumption, increased robustness, or increased speed, provided that the natural language processing modules have been designed in such a way as to optimally support any such trade-off.

4　Linguistic theory and linguistic functionality of language products

Linguistic theory oriented towards the cognitive paradigm assumes as its subject matter the linguistic competence of an "ideal speaker-listener, in a completely homogenous speech-community, who knows his language perfectly and is unaffected by such grammatically irrelevant conditions as memory limitations, distractions, shifts of attention and interest, and errors (random or characteristic) in applying his knowledge of the language in actual performance" (Chomsky 1965:3). This linguistic competence, it is furthermore assumed, can be adequately modelled on functionalist premises within the paradigm of symbolic representation and symbol manipulation as it originated in research on artificial intelligence (for an excellent exposition of the approach see Habel 1986:6 ff.). To the extent, however, that within this orientation the scientific interest in a computational model of language is linguistic competence, and thus more broadly, the human mind and human understanding, it is also assumed as a matter of course that the natural language processing modules draw on a representation of linguistic knowledge that represents linguistic knowledge as general as possible.

From a language technology point of view, the primary goal is not a general, or presumedly cognitively adequate, representation of linguistic knowledge, but rather an optimally engineered solution to specific, empirically validated, problems in the area of interface or autonomous tasks systems. Here, of course, generality of the linguistic descriptions also is an issue in order to avoid *ad hoc* representations. But although there may be a software engineering level of general representation of linguistic knowledge as indicated above, the main problem for an application oriented linguistic theory is rather to provide a theoretical basis for optimizing trade-offs between different linguistic and non-linguistic requirements for a particular application (cf. Obermeier 1989: 235 and 100).

In detail, one problem is that users may not require 100% recognition, but instead expect performance improvements with respect to other software quality criteria, as was discussed in the context of ergonomic design of language products.

A second problem is that not all linguistic phenomena need to, or can, be assigned the same importance, as is in particular the case in language checking. As an example, let us consider the following list of percentage of errors in the so-called Heidelberg corpus (see figure 6), a corpus of errors in German collected in the course of ESPRIT II project 2315, Translator's Workbench, by the Heidelberg department for Computational Linguistics based on 50 examination papers written by foreign students (Hellwig/Bläsi 1990).

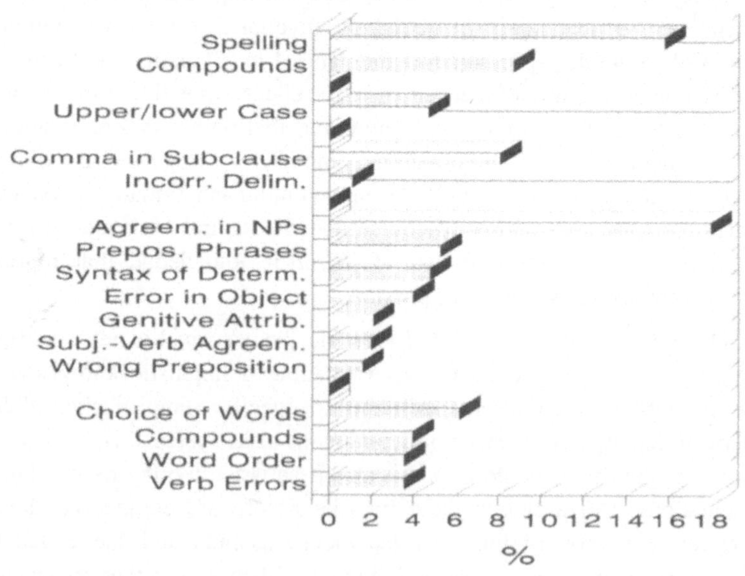

Figure 6

As can be expected, certain errors like spelling, or NP agreement, are more frequent than others, e.g. with respect to word order. It is interesting to note, however, that the most frequent errors also require the least processing efforts. Thus, from the language technology point of view, we clearly have an indication here of a reasonable trade-off between theoretical costs and practical benefits, where for the purposes of a particular application the kind and number of errors dealt with could be changed, if there also were an additional classification of errors with respect to importance to the user.

In order to satisfy these requirements on linguistic theory, what is needed, I propose, are grammar formalism that are *gradable* with respect to the depth and accuracy of the linguistic processing. Thus, given a language L, a degree of permissible errors E, a specification of the required depth of analysis D, and a measure of (polynomial) time T, what is needed is a grammar formalism G such that a deterministic automaton can be constructed on the basis of G that decides within the given set of parameters E, D, and T, whether or not any arbitrary expression is a sentence of L, and that does so in such a way that a relaxation of E or D leads to a calculable reduction of T. While this requirement may be strange to some in the context of computational linguistics, it is in fact very natural, and there are many examples in other fields. In telecommunications engineering, for example, there obviously is a trade-off here between the quality of the transmitted signals, i.e. the reduction of noise, and the amount of information (in Shannon's sense) that can possibly be transmitted simultaneously. Based on empirical findings, the band-width of frequencies to be transmitted over a telephone line nowadays is restricted in most cases to a range between 1500 and 4500 Hz, although the human ear is capable of perceiving a much broader range. Clearly, engineering requirements of the above kind naturally encourage accepting empirically validated restrictions. In contrast to computational linguistics, however, the theoretical foundations of telecommunications engineering allow for an adequate representation of the kind of trade-offs as discussed above.

Trade-offs between accuracy and efficiency are, of course, also a commonplace in computer science. Surprisingly, however, the recent discussion on computational complexity and natural language (Barton/Berwick/Ristad 1987) has had only litte effect on the presentation of formal frameworks in mainstream computational linguistics, although some frameworks in view of their complexity and general behaviour appear to be better suited for representing that trade-off than others (see, for example, Hausser 1992, or Barton/Berwick/Ristad 1987: 256 ff). Present discussion on efficiency always seems to assume from the start a high degree of grammatical and morphological accuracy, and to the extent that it is recognized that a more efficient processing of simple cases may require the relaxation of some general grammatical constraints (such as agreement conditions), it is not clear how the particular processing effort for the simpler cases is in fact reduced.

Without a linguistic theory thus suitable for real applications, language technology can provide us for the time being only with *heuristics* of how to construct natural language products. On the one hand, grammars and parsers need to be evaluated and classified using parameters such as efficiency, expressiveness, completeness, and decidability of the respective formalisms (cf. Wahlster 1989:216). On the other hand, language technology heuristics will also have to include well-founded advice on the design of the whole system, given specific application tasks, ergonomic considerations, hardware and lingware restrictions. Given that such advice will generally be based on experience, real progress will only be made, once language technology can build on a theory of natural language processing that conceives of natural language processing applications not as a transfer from theory to practice, but in its foundations and consequences intrinsically supports the engineering approach to natural language processing.

5   Elements of a natural language processing technology

The very idea of language technology in its present state, no doubt, is still faint and needs to be spelled out in detail. Nevertheless, I think, the problems are getting clearer, and misconceptions can be resolved. By way of conclusion, let us summarize the main elements of a natural language processing technology as it has been discussed above.

The main task of language technology is to find solutions to technical problems in all areas of natural language processing, and to implement them in an optimal way given particular linguistic, computational, and financial constraints. The key factors that play a role in the most efficient design solutions of language products comprise the aspects *software engineering, software ergonomics,* and *linguistic theory*. In addition, individual programmes need to be optimized with respect to at least the main processing parameters *depth* of linguistic processing, *accuracy* in the sense of errors allowed for the recognition process, and *speed* of the recognition process. In order to arrive at the design of a language product, language technology offers various methods and tools. In my estimation, the key element is a careful empirical evaluation of the ergonomic environment and the required linguistic functionality. While language technology at present offers at best a number of heuristics on the combination of different processing strategies at various levels, the interdependence between various parameters may also be provided a theoretical basis, and thus could be exploited for a more precise calculation of trade-offs in the design of natural language products.

In contrast to a computational linguistics mainly oriented towards cognitive linguistics, language technology assumes a different attitude on the side of the user towards the processing of natural language by a computer than by a human. The key difference is the assumption that when human users communicate with a computer, they change to a different language register, so-called *computer-talk* (see Krause & Hitzenberger 1992). The attitude of computer-talk concerns, on the one hand, the users own linguistic behaviour, and, on the other, the tolerance by which they judge the computer's natural language processing. Thus, depending on the application, human users may be prepared to accept a computer natural language processing which does not deal with all linguistic phenomena, or a recognition rate that is significantly below 100%.

Economic and technological development crucially depends on a public appreciation of the technical problems to be solved. Let us not forget, finally, that the markets for natural language products are just developing in that sense: Various kinds of natural language products for the first time gain wider acceptance, and by doing so make users of these products more aware of the linguistic problems involved. In the very next future we will see, I expect, an increasing demand for a higher quality of natural language products that, in turn, will stimulate the development of language technology. However, as natural language products require a very high degree of linguistic literacy and appreciation, in the long run the field of natural language processing applications will be successful only, if the markets show a still higher degree of awareness for linguistic problems and their solutions.

6   References

[1] Carbonell, "Requirements for Robust Natural Language Interfaces: The Language Craft™ and XCALIBUR experiences, Coling: Bonn 1986

[2] Engelien and McBryde, Natural Language Markets: Commercial Strategies, Ovum Ltd: London 1991

[3] Hausser, Complexity in Left-Associative Grammar, Theoretical Computer Science, Vol. 103, 1992

[4] Heid, "Study Eurotra-7: An early outline of the results and conclusions of the Study", Stuttgart 1991

[5] Hellwig/Bläsi, "Theoretical Basis for Efficient Grammar Checkers, Part II: Automatic Error Detection on Correction", ESPRIT Project 2315, Translator's Workbench, Deliverable 1.5: Heidelberg 1990

[6] Heyer, "Probleme und Aufgaben einer angewandten Computerlinguistik", KI 1/90

[7] Heyer/Kese/Oemig/Dudda, "Knowledge Representation and Semantics in a Complex Domain: The UNIX Natural Language Help System GOETHE, Coling: Helsinki 1990, p. 361-363

[8] Heyer/Khatchadourin/Modiano/Waldhör, "Use and importance of Standard in Electronic Dictionaries: The Compilation Approach for Lexical Ressources, Library and Linguistic Computing, VOL. 9, No. 1, Oxford University Press 1994, 55 - 64

[9] Krause, Computer Talk, Olms, Hildesheim 1992

[10] Liebeherr and Specker, Complexity of Partial Satisfaction, Journal of the Association for Computing Machinery Vol. 28 No. 2, 1981

[11] Möller, "GOETHE user study: An empirical evaluation of ergonomic aspects of a natural language user interface", TA TRIUMPH-ADLER AG, Report No. RR-P-91-06: Nürnberg 1991

[12] Obermeier, Natural language processing technologies in artificial intelligence - The science and industry perspective, Ellis Horwood: Chichester 1989

[13] Pahl/Beitz, Konstruktionslehre, Springer: Berlin, Heidelberg, New York 1986

[14] Schnelle, "Beurteilung von Sprachtechnologie und Sprachindustrie", Colloquium Sprachtechnologie und Praxis der maschinellen Sprachverarbeitung, Reimers-Stiftung: Bad Homburg 1991

[15] Wahlster, "Zum Fortschritt in der Computerlinguistik", in: Bátori, Hahn, Pinkal, Wahlster (Hg.), Computerlinguistik und ihre theoretischen Grundlagen, Springer: Berlin, Heidelberg, New York 1989

Software Ergonomics of Natural Language Systems

Magdalena Zoeppritz
IBM Germany Science Center
Institute for Knowledge Based Systems
Computational Linguistics
Tiergartenstraße 15, D-69121 Heidelberg

1   Introduction

Development of natural language systems, apart from the research interest, always worked on the assumption that such systems would be useful in practice. With the arrival of systems and prototypes that can be used, the question needs to be considered as to what makes them usable and what detracts from usability. The initial idea, that the usability of natural language systems was dependent solely on their linguistic capabilities, turned out to be mistaken. Like other systems, they need reasonable diagnostics, legible manuals, and must fit into their environment. As natural language systems, they differ from other systems in how they foreground specific facets of human language behaviour.

I shall try in this paper to present initial results - from systematic studies as well as more anecdotal observation - on the use and development of natural language systems from a software ergonomics point of view. This view differs from a purely linguistic view in that the linguistic capabilities are not of interest by themselves, but because they have an impact on usability and that on the other hand user tasks and expectations, system environment and other ways of performing the required tasks are also taken into account.

Natural language systems can be roughly grouped by purpose into:

- Query language
- Command language (including intelligent help)
- Machine translation
- Speech to text (or query)

- Text (or answer) to speech

- Text critiqueing (advice on grammaticality and style)

- Content scanning (from filtering mail by subject to abstracting)

The following sections will mainly deal with natural language data base query. The observations from query carry over to some extent to natural language commands, because there also the 'addressee' is a machine, but the problems are certainly different for systems with human readers/listeners. Some of the observations concerning screen layout and integration may apply generally, but suggestions for implementation in other than query systems are outside my competence.

2   Methods for studying natural language systems

The methods that have been used in studying natural language systems shall be only briefly outlined here. More detailed discussion is found in Zoeppritz (1986).

2.1   Simulation

Subjects enter natural language questions, system reactions are simulated by an experimentor playing the role of the system. (Malhotra 1975, Kelley 1984). This method is very useful for collecting requirements for systems, because subjects are not confronted with the restrictions any actual implementation will have. If simulation is used to find out whether restrictions can be tolerated, or which restrictions are possible, simulation leads to unreliable results. It is virtually impossible for a person to react consistently within a given restriction; as a result, the amount of effort required to implement what has been simulated - if it can be implemented at all - is easily underestimated. Similar problems occur with rapid prototyping, where the feasibility of full-scale implementation of the features suggested by the prototype can easily be misjudged (Seaton/Stewart, 1992).

2.2   Comparison of systems

Comparing different systems in controlled experiments has the advantage of a stable environment. Specific tasks are set and shall be performed with either system, performance is measured according to predefined criteria. Since such tasks are normally communicated in natural language, there is an added problem with natural language systems: how can the tasks be presented without influencing the formulation of questions (Ogden/Brooks 1983). A general problem of this method is the choice of a frame of reference - one needs to know beforehand which factors are relevant and test these factors as cleanly as possible. This requires a lot of resources and it is often the case that the scope of the experiment gets so narrow that the points where the systems differ remain outside the test (Reisner, 1981). Another problem is how results from controlled experiments are used - the status of statements like 'system a is better than system b'. Though the method in fact does not permit any conclusion about the absolute merit of either of the systems compared - both may be bad - this is easily overlooked when results are reported.

2.3   User studies

Field tests with users are well suited for observing actual user behaviour and are closest to the working situation. The questions come from actual information needs and not from tasks set for the purpose of the experiment. System restrictions can be viewed in relation to the intended task. However, feedback from the system influences user input. Since users are interested in getting their work done, they tend to use questions that work. While this has been viewed as a disadvantage of the experimental setting, it is actually an advantage if one considers that perfect natural language systems are out of reach: it is then necessary to know whether one can get used to the characteristics of imperfect systems and whether the benefits outweigh the cost. Field studies are essential for learning which factors, both on the user's side and on the system's side, play a role in how natural language systems are used. The real problems lie in evaluating and assessing the meaning of observations made in field studies. Questions as to the importance of observed phenomena can usually not be answered by statistics on occurrence in the test. Though predictions about future occurrences are possible if observed phenomena are related to working situation and system architecture, such predictions can be discounted as speculative. Lately, there seems to be a change in paradigm, or at least a loosening of the paradigm prevailing in the past, so that there is hope that results from field studies will in the future be taken seriously in the scientific community also (cf. Shneiderman/Carroll 1988).

2.4   Problems for Methodology

Investigating natural language systems with conventional methods for studying usability is complicated by the fact that the **preconditions** under which natural language systems operate are quite different from those of other systems:

- Users know the language better than the system does:

  Any system is insufficient with respect to the user's command of the language. A change can be expected in the future to the effect that natural language systems are compared to other natural language systems and not primarily to the human command of natural language.

- Natural language, at least initially, is associated by some users with the 'real world' and not with the system world in which it operates:

  So it could be observed several times that answers from the data base were interpreted as answers about the real world, not as answers with respect to data base content, with the side-effect that insufficient or incorrect data were taken as indicators of malfunction.

- There are 'folklore' assumptions about how such systems function (e.g. that only certain keywords are analyzed); such assumptions can impede efficient use of the system.

Observation shows that these conditions play a role in the use of natural language systems, but it does not seem to be clear yet how they can be taken into account in systematic studies of use.

3   Summary of results

The bulk of results from user and laboratory studies of natural language data base query systems come from several studies with two versions of one system. The results as well as results from previous studies with other systems are discussed under various aspects in Krause (1982), Zoeppritz (1983), and Jarke/Krause/Vassilou (1986). Details on methods and findings are found in these reports and the references cited there, so I shall just briefly summarize the main points:

- Users do not communicate with systems in the way they would with a person. Several features of observed input are specific to communication with computers:

> Input is careful, there are fewer typing errors and less grammatical errors than occur in person-to-person communication.

- There are differences in the use of constructions. These phenomena have been followed up and investigated in detail by Jürgen Krause (cf. Krause/Hitzenberger 1992).

Users adapt to the limitations of a natural language system, the more easily the more the required functionality is there and consistently implemented.

The effect of restrictions differs according to type of restriction:

- Small vocabularies that fit the application domain do not cause problems.

- Restrictions in syntactic and syntacto-semantic variation (different ways of saying the same thing) are the most irritating.

- Semantic restrictions always mean restriction in functionality. The extent to which they are tolerated depends on the need for the functionality in question.

There are unexpected dependencies among features that need to be explored further. Formatting, calculation, coordination are expected features, because they are (or, as *and* and *or*, seem to be) available in other computer systems and languages. Absence of these features discourages exploration of features that exist, e.g. relative clauses.

Learning is needed for natural language systems as well, though of a different form.

The complexity of natural language questions remains relatively stable over time and user, this also means that people may begin with - in formal terms - rather complex queries, i.e. queries whose equivalents in the query language go beyond what non-expert users of query languages use and may go beyond what is taught in query language courses.

People suspect errors in the system rather than in the data or in their handling of it. Typing errors are overlooked, missing data are not believed until repeated questions bring confirmation. This finding may be specific to systems that 'admit' that they cannot handle everything, because with other systems it could be observed that even the most glaring misinterpretations went undetected by users (a form of ELIZA-effect).

There are advantages to using natural language:

- Users can ask questions according to their information needs with little knowledge of the specifics of the data base.

- They feel certain that their questions correspond to what they intend to ask (as opposed to 'debugging' formal language expressions).

- Natural language expressions are generally more concise than their formal language counterparts.

The studies with USL were conducted more than ten years ago. In the meantime, both systems, users, and the environment have changed. Systems have become much more sophisticated, today's naive users are much less 'naive', and the system environment includes features like windows, mouse, menus, etc. that did not figure in the studies then.

While the basic findings summarized above still hold, and can be confirmed by observing use today, some of the usability issues raised have become clearer and new issues have come about.

4   Usability, habitability, and consistency

The observations show that natural language query systems are **usable**, even if their coverage and functionality is restricted not only with respect to human language but also with respect to the tasks to be performed.

If missing functionality can be obtained by using a formal query language, it is possible to envisage users who will switch from one language to the other, using each for what the other will not provide, but this path will be open only to experienced users of query languages and it will be added strain for them. Furthermore, though experienced SQL-users in our experience have been impressed by how easily complex queries can be correctly generated via natural language, the idea of switching between languages for reasons of functionality does not appeal to them. Therefore, usability very much depends on functionality.

Within the functionality provided, it was found that gaps in coverage were most irritating. It proved difficult to remember which way of asking for a specific function would work and which would not. This difficulty was already seen by Watt (1968). He proposed **habitability** as a necessary feature of systems (see

Krause 1982 for a discussion of this and similar concepts) and tried to define it for English mainly in syntactic terms. Applied to data base query systems, habitable systems would cover the normal ways to address the functionality provided in a consistent way.

The word 'normal' resists definition somewhat but is often clear in observational terms: If users cannot remember which of the ways of addressing existing functionality work and which do not, this shows that both ways are 'normal' and need to be covered. Note however, that this is knowledge after the fact, while as developers we would like to be able to predict.

In his article "On natural language based computer systems" (Petrick 1976), Stanley Petrick critizised a particular style of implementation and in doing so drew attention to an important part of human language behavior, namely **extrapolation:**

> *...the consideration of a few examples usually leads to unwarranted extrapolation as to system capabilities. This is natural because the reader of a paper who observes sample sentences in which conjunction, negation, and quantification occur, for example, assumes that these phenomena can be successfully analyzed when they occur in different ways involving no additional lexical items; unfortunately, this assumption is often wrong.*

While the form of extrapolation may be different for linguist readers (coordination and negation are covered) and non-linguist readers (one can use *and* and *not*), extrapolation is a natural faculty of human beings. Given an example, and being told which function the example exemplifies, people usually can make up examples that are similar.

This faculty can be exploited in approaching consistency. Developers can extrapolate the extent of needed coverage - and the test cases - from the specifications, and users can extrapolate from the examples and counterexamples that describe coverage in the manuals.

In the abstract, and from a development point of view, **consistent** coverage means covering both syntax and semantics for given functions as well as for the functions in combination (Petrick, 1976). This is prerequisite, but then there is a communication problem.

What users actually perceive as inconsistency in coverage of existing function, rather than as separate function that is missing, has to do with the extent to which language behavior is consciously accessible and what users see as **same** or **different**.

```
1. Welche Klasse hat die größte Kinderzahl?

2. Welche Klasse ist am größten?

3. Welcher Schriftsteller ist am größten?[1]
```

The difference between 1 and 2 on the one hand and 3 on the other is not equally obvious to all beginning users, but it can be successfully explained. This would not be true in case of 4 and 5:

```
4. Welche Klasse ist am größten?

5. Welches ist die größte Klasse?[2]
```

A borderline case for explanation is exemplified by the pair 6 and 7, where explanation is possible, but successful adaptation is unlikely (so better cover both):

```
6. Wer verdient mehr als 500 DM

7. Wer verdient mehr als Paul[3]
```

Again, while it is relatively easy to think of individual examples for consistency or inconsistency, or to enumerate problems encountered, there is no way to be sure that one has thought of all possible inconsistencies and one is left for the time being with systematic accumulation and the hope that the ragged edges of present systems will be rounded out in time.

Functionality that can be described in terms of what is consciously accessible can also be communicated, and users can adapt their expectations accordingly. Where semantic restrictions can be stated in terms of functionality, they do not have the impact on involuntary use that was observed for syntactic restrictions (Krause 1982). But if there are too many exceptions in the way existing functionality can be used, whether the exceptions are stated or experienced,

---

[1] *1. Which class has the largest number of children*
2. Which class is the largest
3. Which writer is the greatest

[2] *4. Which class is the largest*
5. Which is the largest class

[3] *6. Who earns more than 500 DM*
7. Who earns more than Paul

users will have the same difficulties feeling comfortable with the system as they have with syntactic restrictions.[4]

4.1   The user's view

The promise of using natural language can lead to overly high expectations. A question like *where do most of our profits come from* may not be answerable from the data alone, but requires judgement, which does not come automatically with the language.

Successful interaction can lead to **overestimating** the system and attributing cognitive abilities to it that are not there (cf. the well known example of ELIZA). Such effects disappear with use, but users should be warned early, in the documentation as well as during demonstrations, to avoid disappointment.

In view of this, there is a need for **fair demonstrations**. They should exhibit only what the system can do consistently. Fairness is even more important for story boards: Where the system is simulated, the onlookers are almost in the position of users, but cannot test their extrapolations by asking their own questions (I owe this observation to Jane Banwart). It is always possible to create query sequences which, in that sequence, imply much more than what is actually there.

Unsuccessful interaction can lead to **underestimating** the system, and, as noted above, this may not only mean avoiding the function that was unsuccessful, but may also extend to at first glance unrelated functions (like not expecting complex clauses from a system that does not cover formatting commands).

The advantages of natural language can be observed at almost all demonstrations: people formulate questions against a data base they have never seen before and criticize whatever does not work (irrespective of whether it can reasonably be expected). This is something they would have no chance of even trying if it were not natural language. Nevertheless, people are not really impressed. It seems that natural language understanding is taken so much for granted that it functions as a **dissatisfier**, i.e. what works is taken for granted, only what does not work is noted. This seems to be in contrast to speech

---

[4]   The impact of inconsistencies may be less severe for translation systems, but should be taken into account when proposing requirements to 'write for translatability' and penalties for not doing so.

recognition, or so it appeared when LanguageAccess was demonstrated at the CEBIT 1992 together with a prototype speech recognition system (Walch et al. 1989, Keppel et al. 1991). To most of the visitors, the transformation of speech to writing was the actual sensation and only few of them found it at all surprising that the system could also understand what was written.

Natural language still has the image of being something for naive users, hence there is no status involved in using a natural language system. But that may be changing. Experienced users of SQL begin to discover that automatically generating SQL-statements via natural language can save much of the time otherwise needed for generating and debugging the complex SQL statements they want to use.[5]

5   Integration into the environment

Natural language does not solve all problems, so, naturally, the way natural language systems are integrated into the working environment, side by side with other ways of interaction play a role in usability. Readily accessing natural language systems as functions among other functions, easily switching between functions, and exporting or importing results is useful.

Other requests go in the direction of multimedia: Graphical interfaces seem to be expected. If written input is possible, then speech should be possible as well.

The best ways to bring different functions together, and how different media can productively interact require further user studies.

5.1   Layout

User interfaces of natural language systems are built along the lines of user interfaces in general. As such they have Color, Function Keys, Icons, Windows, and whatever else is regarded as standard equipment.

I am not aware of studies that explore to what extent this standard equipment, or the way it is usually employed, is adequate for natural language systems.

---

[5] Many information systems departments now provide menu-driven query generation for their end users. While the earlier studies compared writing questions in natural language to writing query language statements, the alternatives for end users now seem to be natural language versus guided query. I am not aware of studies that investigate how formulating in natural language compares to navigation through menus and other forms of guided query generation. While it can be assumed that the results would favor natural language also, such studies would provide detail as to what exactly the differences are.

An empty screen containing just user questions and system answers, as back in the seventies, may not be very good, but a screen full of options, pull down menus, and function keys can be irritating for users who have a specific task in mind and want to keep track of where they are with respect to that task.

The issues then are: what is the optimal ratio between dialog or dialog history and other items shown on the screen? What is the role of screen layout and options or menus in the usability of natural language systems?

At the moment the tendency seems rather to go in the direction of more equipment and larger screens, and flipping among screens and pull-downs certainly makes demonstrations much more impressive. But there could be a conflict of purpose where systems intended also for people who are not overly interested in computers have user interfaces that need the skills of the computer-literate.

6 Learning

One of the advantages expected of natural language systems was that there would be no learning effort required, since people already know natural language. But there still are things to be learned: about the language that the system can handle, about customization, and - for people new to computers - about systems in general. For experienced computer users there is a problem of changing habits.

6.1 Different: language in the system

Natural language systems react systematically in ways that are surprising at first, for instance words are interpreted as defined even when they are used in a different sense, there are restrictions with respect to human language, and there may be inconsistencies that are idiosyncratic to the system used.

6.2 New: Customization

Transportable systems, that are not custom-made for one specific application, require customization before they can be used. This means adapting the systems to a specific data base, adding vocabulary and morphological, semantic and pragmatic information as required for processing.

Adaptation work will not necessarily become less as systems become more powerful. An example for illustration: Customization for comparison on numbers requires mapping words in some way to columns in the data base. If there is a word 'area', mapped to the area of a country, then the word can be used in numeric comparison, e.g. *which area is largest*. If questions like *which is the largest country* shall be supported, additional information is needed on what makes countries large (population figure, gross national product, number of doctors?), which must be supplied depending on available data and purpose of the data base during customization.[6]

6.3   Relearning: asking questions

Simply asking questions seems to be all the more difficult for people the more educated they are about software systems and data bases.

In earlier years it was procedural habits or expectations that interfered:

Please get ID of companies and individuals that have donated more than 20000 in 1981 from the donations table. Take the ID and match up with the alumni or company from the personal information of appropriate tables. List last name, City, State, and Zip of both alumni and companies. (Vassiliou et al. 1983).

Instead of:

Which alumni donated more than 20000 in 1981.

Now it is the expectation of being led through menus or designing menus. In any case, the habit of first translating whatever is wanted into some other form gets in the way of straightforward interaction with a natural language system. If it then happens that the first question does not work for some reason, all fears seem to be justified.

---

[6]   The manner in which this is done best, whether explicitly or by selecting from a choice or by accepting or overriding defaults, is still an open issue. While correspondences like those between 'old/age' or 'expensive/price' can be generalized over applications and therefore made part of the system, default assumptions about how 'large' and 'country' are related may easily be wrong in the specific case and therefore possibly more misleading than helpful in a general purpose system (an interesting account of customization is found in Walker/Nelson/Stenton 1992).

7 Documentation

Documentation for natural language systems has the dual function of informing about the system and about the linguistic capabilities. While the systems aspects: meaning and use of keys, facilities accessible via pull-down menus, navigation among panels, etc. pose similar problems as for other systems, informing about the language is a different type of task than explaining a formal language.

When getting to know a natural language system, one does not get to know the language and the system at the same time.[7] One **knows** the language, i.e. one uses its capabilities without being conscious in detail of what they are and how one makes use of them. Compared to this knowledge, the capabilities of a natural language system are limited in ways specific to that system.

A major function of documentation is then to assist users in developing a feeling for the **outlines** of the system, the capabilities and limitations of the system at hand, so that they can use it with confidence. Exhaustive description is not possible in finite time and on finite amounts of paper, and might not serve the stated purpose if it were possible. But, if we interpret Petrick (1976) and Krause (1982) correctly, exhaustive description may not be necessary. Describing the functionality provided and giving examples of how to address the features (preferably with counterexamples showing the limits) could be sufficient. Users knowledge of the language will let them extrapolate from the examples without extra instruction. Experimenting with the system will either confirm their extrapolations or, where things do not work, extrapolation from counterexamples in the documentation should guide their understanding of why they went beyond what the system can do.

There are two obstacles: First, giving examples of what does not work is needed to find out where exactly the limitations are in concrete rather than abstract terms. But writing these things into a manual may not be good advertising. Secondly, if there are minor inconsistencies in the system, they may not find their way into the documentation for fear of confusing people into using a much too limited language. These obstacles will become smaller the better we know how to build habitable systems.

---

[7] This may also be true e.g. when using a formal language with a new compiler, with the difference that formal languages are much smaller in scope and all of their syntax and semantics are acquired consciously.

8   System response

8.1   Feedback

Providing feedback as to the results of language processing through the system is necessary for at least two reasons: to give users an opportunity to check on the system and to permit choice in case of several possible interpretations of the input (ambiguity).

If the goal of language processing is translation into another natural language, feedback and results of processing more or less go together.

Where the target is anything else - an answer from a data base system or an action to be performed - seeing the answer or the action is not enough to make users feel safe. Additional feedback is needed. This can be given in different ways, among them feedback in the target language or feedback by paraphrasing in natural language the internal representation at a suitable stage.

8.1.1   Target language

The result can be shown in the target language - in case of data base query, for instance, in the target query language, or in a formal command language, if that is the target. This appears contradictory at first, since natural language promises to relieve users from formal languages, but it has the advantage of really making clear what the system did with the user's input. If one realizes that there is something like the difference between active and passive knowledge of formal languages as well, this approach begins to look feasible. The active knowledge of the language may not be sufficient to formulate a query or a command with certainty, quickly, and without repeated attempts, while the passive knowledge may be quite sufficient for checking the plausibility of an automatically generated query or command. If questions - i.e. their formal equivalents - get complicated, feedback in the form of target language results becomes self-defeating. Users are presented with something, but they cannot judge whether it represents their intentions. While the target language has disadvantages as the only kind of feedback, target language expressions can be used for other purposes as well, so they should always be available for those who want to see them.

8.1.2 Paraphrase

Natural language paraphrases of the results of language processing by the system must fulfil several requirements in order to be useful.

Paraphrases should be as explicit as necessary to clearly reflect both the interpretations in question and the differences between interpretations.

They should not have more than one reading (in the context of the application).

Two different interpretations should not result in the same paraphrase.

They should not mislead users into chosing an interpretation they did not intend.

Their preferred reading should not reflect a meaning different from the interpretation paraphrased.[8]

They should be legible, i.e. neither too long nor too garbled to check their correctness or to chose between them.

These requirements are not easy to meet - the less so the larger the range of possible expressions - and the requirements conflict with one another and with the economy of languages. Normally the most explicit and unambiguous way of saying something is not the most natural way of saying it.

Furthermore there is the problem of **convergence**. People tend to adapt their linguistic habits to the communication partner. Adaptive behavior when interacting with a machine takes specific forms and differs from adaptation in human communication, but among the specifics of adaptation to machines are, on the one hand, a tendency to view the system as an authority and, on the

---

8   The following example illustrates this point:

```
Question:     Exportiert Spanien          (does Spain export)
Paraphrase:   Führt der Exporteur Spanien aus
              Generated reading: Does the exporter Spain export
              Preferred reading: Does the exporter take Spain out
Correction:   Führt der Exporteur Spanien etwas aus
              (Does the exporter Spain export something)
```

By explicating the constant *Spanien* into *Exporteur Spanien* the possibility of reading *Exporteur Spanien* as two noun phrases - subject and object - is introduced which, in turn, forces another reading of the verb: *take out* instead of *export*, and that reading is preferred, so the paraphrase is misleading as to the interpretation from which it is generated. To avoid such effects, objects must be added in the paraphrase that are unnecessary in the question and are not present in the representation to be paraphrased. The example also illustrates that such effects, where they occur and how they can be avoided, are largely language specific, having to do with things like word order, gender (determines e.g. possible distance between relative clause and head), or morphology (morphological distinctness allows for more flexibility, neutralization is a source of competing readings).

other, a tendency to make things easier for the system. Both tendencies can be observed in users reaction to paraphrases. If users see their input reformulated, they sometimes take this as correction. Even if they do not like the formulation they see, they believe that the paraphrase represents what they should have said - either because the system would have handled the paraphrase more easily or because the paraphrase formulation is more correct than their input.

While it is possible and necessary to tell users that the paraphrases are explicit and sometimes not very natural for reasons of clarity and that they may contain redundancies that make them inefficient input for processing, it would be unrealistic to expect users not to slip now and then into the linguistic forms that are fed back to them. It would be attractive to think of paraphrases as the language of convergence, as showing the language that the system will process most easily. Since this is certainly in conflict with clarity and explicitness on the one hand, and naturalness of input on the other, this cannot be required or realized. But in view of convergence effects, it is highly desirable that paraphrase-style questions, if they occur, can be handled by the system. As a consequence, the language generated should not be too far away from the language analyzed.[9] Even this much is more difficult to realize than it may seem.

While there are several systems that paraphrase user input, and reasonable ways have been found in them of resolving the conflict among the different requirements, the basic problems underlying paraphrasing, inasmuch as they are different from the problems of text or sentence generation, are not yet well understood.

8.1.3 Clarification dialog

Originally, clarification dialog was intended as an aid in natural language analysis (Codd, 1974). As such it has proved unhelpful (Codd, 1978). What remains of it has evolved into paraphrase and/or menu selection at choice points.

8.2 Errors

Traditionally, errors are of two kinds: system errors or user errors. Among the first are bugs, incomplete implementation of the design or missing caveats in

---

[9] Convergence effects might be less if paraphrases do not paraphrase the input but the expected answer: *the X that are Y*, rather than *Which X are Y*. I owe this suggestion to Olivier Winghart (personal communication).

the manuals. User errors are typing errors, syntax errors, or logical errors of some type.

With natural language systems, a new variety of error is introduced, while the traditional category of 'syntax error' is more in the nature of typing errors (two articles instead of one, remaining parts of a previous query, etc.). The definition of 'correct input' is given by the natural language in question, not by the definitions implemented in the parser. Discrepancies between 'correct input' as seen by the user and 'correct input' with respect to what the system can handle cannot be resolved by asking users to improve their mastery of the language, since the user is not at fault. Improving the system as much as possible - while desirable - is only part of the answer.

Since all implementations of natural language in a system will remain incomplete, discrepancies between user input and capabilities of the system must be reckoned with and ways need to be found to deal with them, preferably without frustration or incorrect results.

True input errors, like typing errors, are easily overlooked. To counteract that, marking possible typing errors - signalling words that are not recognized - has proved useful. Unfortunately, this does not catch word repetition and similar errors.

Other mismatches between input and system reactions fall into the area of discrepancies. There may be remedies, as in the case of incomplete customization, or the user has encountered a limitation of the system. Proper diagnostics should help users to decide what the problem is.

8.3 Tolerance of errors

Typing errors could also be corrected automatically, but how useful that is depends on the type of system. Trivially: an overnight batch job should not simply stop if an error is found; on the other hand, an interactive job should not spend extended time on correcting an error that users may have already spotted as they hit the Enter key.

In an interactive environment, correcting typing errors may be costly, involving string manipulation and a search space as large as the entire data base. Marking possible typing errors was found more useful than not marking them, even though - as in our case - correctly spelled names were marked, while repetition errors were not marked.

Correcting errors other than typing errors presupposes the same type of information as good diagnostics do (cf. below) and additional processing for correction. There do not seem to be feasible proposals with a tolerable ratio of correction proper to false alarm. Processing faulty input without confirmed correction appears dangerous to me, it may easily lead to incorrect results and make the entire system unreliable.

In view of the fact that faulty input does not occur very often in query (Krause, 1982), dealing with it does not seem to be a major factor in the usability of natural language query systems.

If certain constructions appear regularly, though they are ungrammatical under conservative judgments of grammaticality, the best 'tolerance' is not regarding them as errors but as variants to be included. This may even extend to variants that are outside normal usage. There is strong evidence of an emerging style or register that is used specifically when interacting with machines (Zoeppritz 1985, Krause/Hitzenberger 1992) which interactive systems should try to accommodate.[10]

8.4   Diagnostics

If the system cannot handle user input, there will be an error message. Ideally the error message should be explicit enough for users to know what to do next. It follows from the expressiveness of natural language that there will have to be more error messages than there need to be for a formal query language. And the fact that the language implemented is a subset of the user's language means that error messages should have something to say about that difference. This amounts to saying that the system needs to know what it is that it cannot do. This is one of the biggest open issues: what **in** the system can indicate which feature **outside** the system is addressed by some input? This looks like a logically unsolvable problem but is mitigated by the fact that the developers know the difference between the language and the implementation: diagnostics then are the vehicle to communicate this knowledge. The extent to which this can be done depends on the architecture of the system.

If the architecture of the system allows distinguishing between *cannot parse*, different customization problems like *definitions do not allow saying 'X works*

---

[10] While the task of data base query restricts coverage to the extent that many features found e.g. in expository prose can be left out, accommodating the features of computertalk amounts to an **extension** rather than a reduction of coverage.

for Y', different semantic failures or data type failures like *cannot do comparison on a character field*, this is definitely more helpful than *not understood, please rephrase.*

While architectures that force one analysis run into the danger of selecting the wrong one, and have by and large been dropped, architectures that pursue multiple analyses have the problem of selecting from several error messages the one error message that corresponds to the intended reading of the input. I know of no proposal that would solve this problem.

An architecture that uses syntax not just for parsing but also for directing the interpretation permits quite specific diagnostics for failure to interpret, once input has been parsed, but may have no criteria for spotting the exact cause for failure to parse.[11] In situations of failure to parse - or failure to interpret for non-obvious reasons, Jane Banwart (personal communication) observed that users found an option very helpful that let them see the words of the input and their definitions: whether they came from the general vocabulary or the customization and which category had been defined or - in case of unknown words - assumed. Options that help users diagnose for themselves are an interesting complement to the - probably never ideal - diagnostics from the system.

Assuming diagnostics are available inside the system, correct and at a sufficient level of detail, there is still the question of how to formulate them so that they advise rather than reprimand the users who see them.

9   Ergonomic aspects in the development process

Natural language systems get large and they need to represent a lot of phenomena that are by nature complex and often appear idiosyncratic at first glance. Such systems cannot be made elegant and simple by reducing complexity. Any reduction introduces gaps in coverage and functionality that make the system seem unpredictable to the user. On the other hand, maintenance and extensibility require modularization and uniformity of design.

---

[11] Proposals along the lines of 'longest possible match', i.e. assuming that the cause of failure would most often be lack of a rule bridging the gap between the largest and the smallest adjacent subtrees did not work out. Looking at many failures in context leads me to suspect that the individual subtrees carry information that might be used to find the breaking point in each individual case, but how this could be exploited and whether it can be done at all efficiently needs further investigation. The problem stated here in terms of rules and subtrees is not confined to this type of architecture, though the details may differ in other architectures.

This can be achieved only if the design leaves room for variation, i.e. not uniformity in a strict sense, but harmony at a rather high level of complexity.

In view of this, suitable development tools are vital. A useful overview of issues and proposals in the area of grammar development are found in Erbach/Uszkoreit (1990).

9.1   Development tools

Developers are users as well. They need reasonable diagnostics, suitable programming languages, and transparent representations to assist them in their work. None of these requirements are new, so a summary is sufficient:

- **Programming languages and debuggers with reasonable diagnostics**
  If locating a minor error during implementation takes too much time, it is likely that some important part of the implementation idea gets forgotten.

- **Expressive formalisms**
  This requirement is to a certain extent in conflict with attempts to reduce the power of formalisms. The latter can often be achieved only at the cost of expressiveness, which is undesirable. But of course, powerful languages can be abused.

- **Transparent representations**
  Under transparency I would also include means to make developers aware of what they are doing. The more high-level the language, the more likely it is that efficiency issues are hidden from the (developer-)users. While it should be up to the developers to choose the most suitable formulation, knowing about efficiency is useful when there are alternatives.

- **Color, windows, cut and paste, etc.**
  Such things are more or less standard now, but their contribution is minimal, unless the requirements above have been met. In that sense they are very desirable additions, but cannot substitute for any of the above.

While the requirements may be clear, it is not at all clear what specific tools, debugging aids, properties of formalisms etc. will actually contribute to meeting them.

9.2　Methods for testing and evaluation

9.2.1　Testing natural language systems

Natural language systems cannot be tested exhaustively, i.e. all instructions in all combinations cannot be tested. And in any case there will be extensions as a consequence of adaptation to user environments. We have found useful a combination of **systematic testing** of each of the parts together with **sample testing** of combinations.

A part in this sense would be e.g. the interpretation of sentence types with verbs taking different arguments in all their sequences, or the interpretation of complex noun phrases. When systematically testing verbs, the nominal elements need to be different only in as far as the verb rules refer to these differences. Similarly, the verbs can remain constant in the tests for noun phrases.

We use constructed sentences for systematic testing with the advantage of control over vocabulary and construction types (Rare constructions, which are necessary for systematic reasons, may be difficult to find in natural contexts). The interaction between verbs and complex nounphrases can then be checked through sample testing.

Good candidates for inclusion in sample test suites are those inputs from past user studies or demonstrations in which something unexpected or undesirable happened - questions that exhibit side-effects. Such side-effects may be easily explained afterwards, but the fact that they were not anticipated indicates a lack of awareness in the relevant area. Over time a feeling develops for problematic combinations, things that need to go into the sample test right away, and a - much less reliable - feeling for which combinations are unproblematic and can be left out.

Ideal sample test suites test those places where problems are to be expected, depending on language and implementation, and if possible only those places, but I think we are still far from having even good sample tests. All this takes a lot of time, effort, and above all, test cases. The assumption that a few hundred test cases could suffice has proven to be mistaken (see also Banwart/Berrettoni 1992, this volume, in the subject of testing and tools).

9.2.2　Testing tools

Since tests should be run and checked frequently (after each change in the code) in order to detect regression early and facilitate error hunting, it is useful to automate testing procedures as far as possible. If tests can be run over night in batch and if the results can be compared automatically with previous results and changes are marked, this is a big and necessary help.

Tests need to be modular - depending on how the system is constructed - so that results at a specific level can be compared and checked separately. Otherwise regression at one level can be easily overlooked because a change at another level makes all results different.

9.2.3 Transportability/Reusability of test corpora and testing methods

In view of the effort needed for collecting test sentences and writing test programs - work that is done in each natural language project by and large separately, even if more or less intensively - there is the obvious question of reusable test batches and test programs. Indeed, that would be extremely useful. Realisation, however, requires additional effort within projects and cooperation between projects. Cooperation with respect to vocabulary and phenomena to be covered (assuming one can agree on what constitutes a phenomenon) and effort with respect to test programs - general enough for everybody but adaptable to the formalisms and modularizations of each specific system. At this time, there is exchange of test suites between some projects, but test programs are exchanged only where formalisms and goals are the same or sufficiently similar.

For the English language there was a beginning with 'Fehders 99' (a set of 99 sample sentences designed to test query). A large amount of testing materials was presented recently in *Natural Language Sourcebook* (Read et al. 1990). Unfortunately, the coverage is not nearly as complete as the title suggests, large areas of syntactic/semantic variation are not represented and there is far too little discussion of the methods of test case collection.

9.2.4 Comparative tests of different systems

Willée (1983) reports on comparing the performance of different research parsers for German with respect to a single test suite. The problems encountered there increase if the comparison extends to the quality of parses and the results of semantic processing. So currently the Treebank project at the University of Pennsylvania is still comparing syntactic parsers, but not planning to go beyond that (cf. TechMonitor™, SRI 1991).

The comparison of products requires a lot of resources and the actual results of in-depth analyses are usually not published. What is published is too superficial to be reliable, largely based on claims, successful demonstrations, and extrapolation from well-guided hands-on sessions.

One of the problems of evaluation is drawing conclusions from testing without making **attribution errors**. If a sentence testing a specific feature fails, the reason for failure may have nothing to do with the feature tested.[12] Also, if a sentence succeeds that the evaluator regards as difficult, this may not mean that 'easier' variants would have worked.

We do not know how to do 'Warentest'-type quality testing of natural language products yet.

9.3 Internal Evaluation

9.3.1 Consistency of Coverage

Starting from a given set of **what** shall be covered, the next question is **how** that coverage can be realized in order to be useful.

How does one meet requirements like: *The system shall process questions, coordination, negation*? As said above, coverage must include normal ways of using these features, otherwise users may not believe they are there at all. Furthermore, coverage of each feature must be integrated with other features covered, because, if users have to remember a lot of detail about which combinations work and which do not, experience shows that this is not possible. It is more likely that such features are avoided, so use of the system may dwindle down to *List X*, i.e. inconsistently implemented functionality might just as well not be there at all.

This means for instance: if conjunction and negation are covered, this is consistently true only if negation in conjoined phrases is covered as well, and only then will the functionality offered also be used.

Consistency is the result of a process. The initial set of phrases representing the envisaged coverage will most likely be incomplete and will have to be rounded

---

12 Variants of the following questions were used in an internal evaluation to test temporal and locative adverbials: *list modules written in Berlin* and *list modules written when*. Neither sentence worked, and since the evaluator's attention was on adverbials, he reported gaps in the coverage of adverbials, though the actual gap was elsewhere and the same in both sentences (passive relative clauses without pronoun as post modifiers). It is easy even for evaluators experienced in natural language processing to become unaware that they are testing several features at the same time, not just the one that they are focussing on. The example also sheds light on consistency from this user's point of view: absence of adverbials is more likely than absence of this specific construction.

out.[13] Repeated internal reviews are needed to communicate the gaps discovered and make plans to close them. This may mean fixing a bug, adding to the implementation, or redesigning a portion of the code. While the difficulty of closing a gap should not make a difference to the priorities assigned to closing it, in practice it often does, because naturally the future impact on users is perceived less clearly than the immediate impact on the developers.

There may be borderline cases where consistency in the sense described is not possible given the state of the art. But then such cases need extra documentation and justification and should be kept to a minimum.

Designing for consistency of coverage slows down the development process, according to Petrick (1976). My experience leads me to believe that this is not necessarily the case, because it also reduces the time needed for revision and redesign where consistency requires extensions that were not thought of at first.

Needless to say, test suites need to be continually updated so that inconsistencies, once they have been discovered, do not come back in.

9.3.2 Evaluating the design

Coverage and consistency of coverage can to a large extent be verified by testing whether the proposed coverage actually exists. Evaluating the design of the components: grammars, semantic construction, etc. seems hardly possible at this time. Properties like clarity and elegance can be seen, but whether the code covers as much as it should or is extensible in the required direction may be more difficult to predict. With an unruly subject like natural language there is always the danger of ruling out parts of coverage when streamlining the design, unless this is done very slowly and carefully.

In a system designed for more than one language, adaptability to the different languages provides an additional perspective on evaluation. In our experience there has been much evidence that if a function is difficult or even impossible to implement in one language, then there is a major design flaw in the general linguistic code. The problems will be there for all languages, but may be more apparent in some rather than others.

The consequences of - even the most reasonable looking - design decisions are generally not realized in their entirety by the developers (this is hardly possible,

---

[13] In a multilingual architecture, the rounding out process involves looking at the relevant expressions in all the languages planned for inclusion. What is rare usage in one language may be quite normal usage in another.

otherwise the problems with humans simulating a computer would not exist, cf. above). To discuss design decisions by argument is rather difficult. Each argument of type: "then you cannot ask XXX" can be countered by: "but you can ask YYY to get this information" or "that's not in the specifications", both may be trivially true, since specifications[14] must rely on extrapolation as well, and asking 'YYY' may indeed be successful. The question is, would users naturally think of asking 'YYY' if 'XXX' has failed?

Understanding the weight of such arguments needs user experience which at this time is hardly available, at least publicly. An important source of such experience are demonstrations of the system where not only pretested questions are typed in, but 'questions from the audience' are used as well. If the system to be demonstrated has been developed to any extent beyond the toy stage, the demonstrators cannot know all limitations and their consequences by heart. Then it is highly probable that the demonstrators themselves are confronted with - in the user's view - inconsistent behavior of the system, i.e. they are likely to fall into the traps they themselves set. Such experience can contribute to a state of mind where not only the internal consistency is seen at the time of development but where one can in time better imagine which consequences a given design decision may have for users of the system.

9.3.3 Consistency versus Cost

Implementing the required coverage consistently may make a system too large for the intended environment (this problem was pointed out to me by J. Arz). Reduction is necessary, but possible only at the cost of consistency. Then it would be useful to know where reduction will least impact the users. As yet we have no firm linguistic basis for that. When dealing with text, as in machine translation, statistics on tagged corpora can provide information on frequency of occurrence - where tagged corpora exist. In the case of query or command, we are dealing with a new type of text altogether and there is little basis for prediction. Clearly, if the most 'normal' way of saying something fails, people may not expect that a less normal variant would work, but developer intuition about what is normal is not always reliable. It would be desirable to know exactly why a certain construction can safely be left out while leaving out another will severely impact usability. In our experience, whatever function was offered was also used (cf. Lehmann/Ott/Zoeppritz 1978), with one exception:

---

[14] Cf. Heyer/Figge (1990) for a discussion of specifications.

we still have not had an application using verbs with genitive complements, but this is more likely incidental to the application domains we have been dealing with.

So the open question is: How does one find linguistically justified answers to cost/benefit problems?

10 Summary

Previous experience with developing and first experience with using natural language systems have answered some of the open questions, for instance the principal question of whether there are any advantages to using natural language when interacting with computers. At the same time a lot of new questions have come up that need investigation.

I wish to thank Jane A. Banwart, Olivier Winghart and Gerhard Heyer for comments, suggestions, and encouragement, and Herbèrt Leass for helping me with English.

11 References

[1] Banwart, Jane A., Sandra I. Berrettoni (1992): "A Practical Approach to Testing a Natural Language System: Tools and Procedures". This volume.
[2] Codd, E.F. (1974): "Seven steps to Rendez-vous with the Casual User" In Klimbie and Koffeman (eds): *Data Base Management*. Amsterdam.
[3] Codd, E.F. (1978): "RENDEZVOUS Version 1: An Experimental English-Language Query Formulation System for Casual Users of Relational Data Bases", Report RJ2144, IBM Research Laboratory San Jose CA.
[4] Erbach, Gregor, Hans Uszkoreit (1990): "Grammar Engineering: Problems and Prospects". Report on the Saarbrücken Grammar Engineering Workshop. CLAUS Report 1. July 1990.
[5] Heyer, Gerhard, Udo Figge (1989): "Sprachtechnologie und Praxis der maschinellen Sprachverarbeitung". *Sprache und Datenverarbeitung*. 13.2:41-51.
[6] Jarke, Matthias, Jürgen Krause, Yannis Vassilou (1986): "Studies in the Evaluation of a Domain-Independent Natural Language Query System". In L. Bolc, M. Jarke (eds.): *Cooperative Interfaces to Information Systems*. Heidelberg: Springer. 101-129.
[7] Kelley, J.F. (1984): "An iterative design methodology for user-friendly natural language office information applications."
[8] *ACM Transactions on Office Information Systems*. 2.1:26-41.
[9] Keppel, E., H. Cerf-Danon, S. DeGennaro, M. Ferreti, J. Gonzales (1991): "TANGORA - A Large Vocabulary Speech Recognition System For Five Languages". *EUROSPEECH 91*. Genova, Italy, 24th-26th September 1991. Vol 1, 183.

[10] Krause, J. (1982): *Mensch-Maschine-Interaktion in natürlicher Sprache*. Tübingen: Niemeyer.

[11] Krause, Jürgen, Ludwig Hitzenberger (Eds.) (1992): *Computer Talk*. (Sprache und Computer 12). Hildesheim, Zürich, New York: Georg Olms Verlag.

[12] Lehmann, H., N. Ott, M. Zoeppritz (1978) "User experiments with natural language for data base access". *Proceedings of the 7th International Conference on Computational Linguistics*. Bergen 14th to 18th August 1978.

[13] Malhotra, Ashok (1975): "Design criteria for a knowledge-based English language system for management: An experimental analysis." Report MAC TR-146, based on the author's PhD thesis. Cambridge, Mass.: MIT.

[14] Ogden, William C. and Susan R. Brooks (1983): "Query languages for the casual user: Exploring the middle ground between formal and natural languages." In Ann Janda (ed.): *Human Factors in Computing Systems, Proceedings CHI'83, Boston, December 12-15, 1983*. SIGCHI Bulletin special issue. New York: ACM, 161-165.

[15] Petrick, S.R. (1976): "On natural language base computer systems". *IBM Journal of Research and Development*. 20:314-325

[16] Read, Walter, Michael Dyer, Ava Baker, Patricia Mutch, Frances Butler, Alex Quilici, John Reeves (1990): *Natural Language Sourcebook*. Center for Technology Assessment, UCLA Center for the Study of Evaluation, UCLA Computer Science Department, Artificial Intelligence Measurement System, Project Report 13, AD-A233 005

[17] Reisner, P. (1981): "Human factors studies of database query languages: a survey and assessment". *ACM Computing Surveys*. 13:13-31.

[18] Seaton, Paul, Tom Stewart (1992): "Evolving task oriented systems". Penny Bauersfield, John Bennet, Gene Lynch (eds.):*Striking a Balance*. Proceedings of CHI 1992 (Monterey, California, May 3-7 1992), New York: ACM. 463-469.

[19] Shneiderman, B., J.M. Carroll (1988): "Ecological Studies of Professional Programmers: An Overview". *CACM*. 31.1:1256-1258.

[20] Vassiliou, Y., M. Jarke, E.A. Stohr, J.A. Turner (1983): "Natural Language for Data Base Query: A Laboratory Study". *MIS Quarterly*. 7.4:47-61.

[21] Walch, G., K. Mohr, U. Bandara, J. Kempf, E. Keppel, K. Wothke (1989): "Der IBM Spracherkennungsprototyp TANGORA - Anpassung an die deutsche Sprache". *Proc. 11. DAGM-Symposium, Oktober 1989, Hamburg*. (Informatik Fachberichte 219). Berlin, Heidelberg, New York, Tokyo: Springer. 543-550.

[22] Walker, Marilyn A., Andrew L. Nelson, Phil Stenton (1992): "A Case Study of Natural Language Customization: The Practical Effects of World Knowledge". In Ch. Boitet (ed.): *Proc. of the 15th International Conference on Computational Linguistics*. Nantes 23-28/8/1992.

[23] Watt, Willliam C. (1968): "Habitability". *American Documentation*. 19.3:338-351.

[24] Willée, Gerd (1983): "Über die Vergleichbarkeit von Syntaxparsern". *ALLC Journal*. 4:56-63.

[25] Zoeppritz, M. (1983): "Human factors of a 'natural language' enduser system". In A. Blaser, M. Zoeppritz (eds.): *Enduser systems and their human factors*. (Lecture Notes in Computer Science 150). Berlin, Heidelberg, New York, Tokyo: Springer, 62-93. Shorter German version: "Endbenutzersysteme mit 'natürlicher Sprache' und ihre Human Factors". In H. Balzert (ed.) (1983): *Software-Ergonomie*. (Berichte des German Chapter of the ACM 14). Stuttgart: Teubner. 397-410.

[26] Zoeppritz, M. (1986): "Investigating human factors in natural language data base query". In Jacob L. Mey (ed.): *Language and discourse: Test and protest, A Festschrift for Petr Sgall.* (Linguistic and Literary Studies in Eastern Europe 19). Amsterdam, Philadelphia: Benjamins. 585-605.

[27] .Zoeppritz, M. (1985): "Computer talk?". Heidelberg Scientific Center TN 85.05.

A Practical Approach to Testing a Natural Language System: Tools and Procedures

Jane Anne Banwart and Sandra Inés Berrettoni
IBM Nordic Laboratories
Natural Language Processing Department
Lidingö, Sweden

1   Introduction

Most software development projects follow some type of clearly defined process, refined over many years of practical experience. Different phases of development are carefully tracked and documented according to strict guidelines. Until now the majority of natural language systems have been developed as research projects, that is they have been developed without such rigid conditions. However, now that the market for natural language systems is beginning to open up, there is a growing number of commercially available products. It therefore seems timely to start to discuss how the more traditional approaches to software development can be applied to such technologies. This paper concentrates on one vital stage in any development process, namely the test phase. We will show how current approaches need to be adapted to suit the needs of this type of software and, conversely, how much can be gained by looking at the experiences of testers of more conventional software systems.

2   Type of testing

Standard test procedure for software products includes the following activities:

- Unit test

- Function/component test

- System test

- Performance, robustness, usability

- Code inspections

All these activities are important for the testing of a natural language system and all have different requirements in terms of test cases, tools and personnel. Testing is usually divided into formal and informal varieties. The former requires more rigid documentation of test cases and results, but according to some sources making traditionally informal tests more formal can lead to better results.[1] Another characteristic of formal testing is the use of an independent test team, whereas informal testing is usually handled by the writers of each piece of code. This paper is heavily biased towards formal test procedures, but reference will also be made to the less formal methods of testing. For example, the tools described in this paper are designed primarily for formal testing and in particular for function and component test, but with some minor adjustments could also be used for most other types of test. The discussion on code inspections is mainly relevant to multilingual systems and so can be found in the section devoted to the specific problems of such systems.

We have borrowed two terms from general software testing terminology, that is **white box** and **black box** testing. The former refers to testing where the internal structure of the code is taken into account. The latter refers to testing where only the input and output are considered. Both methods of testing are equally important to guarantee the quality of a system.[2]

2.1 Unit test

Unit testing is usually performed by a developer on his or her own code. The intention of unit testing is to eliminate as many bugs as possible before the modules are combined. This stage of testing is usually not subject to the same constraints as other stages. For example, there is usually no documentation procedure for errors discovered during unit test. In his book on managing software development projects, Whitten (1989) suggests that a formal plan for unit testing can ensure that the testing at this stage is thorough enough. Test cases for unit tests must be as exhaustive as possible for the functions of a particular module. This involves testing large numbers of minimally different test cases. Not all of these test cases will be meaningful inputs to the finished system. For example, when testing a grammar module some of the most useful test cases will be grammatical, but semantically bizarre. Unit testing is crucially 'white box' in nature, since the internal structure of the code determines the test

---

[1] See Whitten (1989).

[2] See Horgan and Mathur (1992) for further discussion of these terms.

cases. There should certainly be a consideration of how useful a test case is to a user in the case of failures,[3] but on the whole that aim of unit testing is to test that the code functions consistently and that there are no unexpected gaps.

2.2 Function test

According to some sources, function testing should be part of the informal test phase. However, for the purposes of this paper it is considered as the first stage of formal testing. This is the first stage at which a number of the modules are tested together. It is especially useful where a group of modules will be combined in a number of different systems. The functional specifications serve as the basis for test case creation. Since the specification is based on user requirements, there is a strong element of 'black box' testing at this stage. However, it is important that the gaps between user-style inputs are well tested. In order to fill in these gaps, a degree of detailed knowledge about the internal design of the system is required. At this level of testing, most test cases are likely to be possible inputs from a user, but since the limits of the system must be tested they will not all be grammatical or meaningful. At this stage of testing, all failures of the system are necessarily documented and fixed changes to the code are recorded.

2.3 System test

At some stage all components of a system must be combined and the system must be tested from the perspective of a user, with all user interfaces complete and all documentation available. At this point, the testing is mainly a 'black box' test, since the tester should see the system as a user will see it. Test cases for this stage can be more ad hoc than at other stages, but should be sufficient to offer a high degree of confidence in the reliability of the system. By this stage the natural language parts of the system should have been tested thoroughly and the majority of new errors found will be in the interfaces to the language component. However, if changes are made in a late stage of the testing to a language component then all language components should be retested for regression errors, using the full set of function test cases. In a natural language system the components are so interdependent that a partial or random test can easily miss an important regression error.

---

[3] See Pfleeger (1992) for a discussion of the differences between **error, fault, defect** and **failure**.

2.4   Other types of testing

To prove the usefulness of a system, it is necessary to test every aspect of that system.

For a natural language system, one very important factor is the performance of the system. In many systems, natural language will often be used as an alternative to other more traditional interfaces which are almost certain to be less complex and therefore have faster response times. Natural language systems must be shown to be a viable alternative with respect to performance. For example, no customer would accept a system which takes minutes to process a query that could otherwise be processed in microseconds. In the case of multiuser environments, simultaneous tests are necessary with different numbers of users to ensure that there are no problems when system resources vary. Tests need to show average processing times, as well as maximum times. Averages must be balanced with figures relating to the frequency of particular types of inputs to a system, so that the figures are not biased by some very unexpected inputs.

It must be shown that a system is robust, in that it will not crash or behave erratically whatever the input. Tests of this nature are not so interested in the output as other tests and so can be automated to a much greater extent. The individual modules and also the system as a whole need to undergo this type of test.

The usability of a system is vital to its success. This type of testing is discussed in many papers, for example Zoeppritz (1992) in this volume.

3   Selection of test cases

The choice of test cases is crucial for the success of any test. Different types of test require different types of test cases. These have already been discussed in the sections on types of testing. This section concentrates on the selection of test cases for the function test phase.

Ideally, testing should be a combination of examples from users and those created specifically for the test. This section will suggest some methods of creating suitable test suites. These include some strategies used for creating test cases for other types of software.[4] Often such strategies are ignored, since emphasis is placed on how different natural language systems are compared with other types of software. There are significant differences, but it is important not to ignore or reinvent strategies that have already been tried and tested.

3.1   User test cases

If a system has been exposed to typical users, then information such as user logs should be used in creating test suites. One potential problem with information from test installations, before a product has been formally tested, is that there may well be limitations in the system that could influence the user in his or her choice of inputs to the system. If a particular input does not work then the user will probably give up on that formulation. This will bias the test suite towards what can be done at the point when the user test is carried out, which is likely to be an early version with less functionality than the intended final version. The exact opposite argument can be used against user input gained using **Wizard of Oz** type experiments, such as those described in Malhotra (1975). That is, the user will be lead to expect too much of the system and will use many formulations that are far beyond the coverage of the completed system. Nevertheless all information from users is vital for good testing, provided it is filtered and supplemented adequately by developers and testers.

3.2   Equivalence partitioning

The essence of equivalence partitioning is to define test suites that cover as many of the potential problems as possible, using as few test cases as possible. The aim is to test the most prototypical examples of each phenomenon, ensuring that all relevant variations are included, but excluding test cases that repeat the same function. This is potentially dangerous since it is difficult to avoid excluding some test cases that look equivalent, but which turn out to have crucial differences. However, since natural language is infinite, it is clearly impossible to test all possible inputs and a division of this nature is necessary. For example, it is necessary to test all possible date formats, but the number of test cases including dates should be limited to those which are significantly distinct. The formats *22/3/92* and *22-3-92* are distinct, but to also test the forms

---

[4]   The last three headings in this section were taken from Myers (1979). These were the strategies that seemed most relevant for testing a natural language system. There are many others that should be investigated further, with regard to natural language systems.

23/4/91 and *23-4-91* would be redundant. However, given a system that handles both U.S. and U.K. date formats, the following would be distinct: *3/22/92* and *3-22-92*. Obviously, this is a simple example, not all linguistic constructions or user tasks can be divided up so easily. The intention of avoiding redundancy in testing is, however, vital to the success of the testing.

3.3   Boundary-value analysis

The latter strategy involves looking for the most central examples. Boundary-value analysis focuses on the examples near or outside the known boundaries of coverage. The intention is to ensure that unexpected input cannot force the system to behave in unexpected ways. If a particular input is not covered by the specification, then the system must not fail. Even if the correct or expected output cannot be obtained, the user must not be mislead or have to deal with a system crash. For a natural language system, typical boundary-value test cases would be the most complex formulations of, for example, questions. The system must either handle the test case successfully or at least give back an error message within a reasonable length of time. If, for example, the conjunction of two nouns works fine, then the system should be tested with the conjunction of three, four or even larger numbers of nouns. The system must also be tested with unexpected inputs, such as ungrammatical or nonsense questions to ensure that the code functions normally in abnormal situations.

3.4   Error guessing

Testing will always rely to some extent on the intuitions of the test designers. Since not all functions can be tested fully, the tester must guess which areas could be problematic and concentrate testing on those areas. This concept is discussed further in the section on the particular problems faced by multilingual systems, in particular the discussion on code inspections. Some linguistic constructions are inherently difficult to handle in any system. Knowledge of this kind can come from either developers or from other experts. Such testing is somewhat ad hoc, but can prove very useful for locating errors that other methods miss.

4   Analysis of results

Testing any system presupposes that testers have a clear idea of what the system is supposed to do (the specification) and also of what behavior will constitute a failure to a user. For a natural language system the distinction between failures and expected behavior is not always clear. This is true of all systems to some extent, but is more apparent for a natural language system. Since the input is natural language, it is often very difficult to draw boundaries in the coverage that are as meaningful to users as they are to developers. Some cases of failure will be uncontroversial, but others will require discussion, reference to usability tests and even changes to product documentation. We have classified the different types of potential failures as follows:

- Badly formed output

- Wrong answers

- Unexpected output

- Too much output

- No output (error messages, diagnostics)

- Unexpected input

4.1 Badly-formed output

Any output that is badly-formed constitutes a failure. For a system where natural language is converted to a command or a query, any syntax error in that command or query must be classified as a failure. Likewise for a natural language output, any ungrammatical or nonsense statements are also failures. Natural language outputs that are ambiguous or difficult to understand are more difficult to judge and decisions about these types of test cases should be taken in consultation with reference to user needs and should not be taken by developers alone. The judgement of developers, in such cases, is often clouded by the knowledge of the design of the system. An output which can be justified by a developer in terms of logical reasoning may be completely incomprehensible to the average user.

4.2 Wrong data produced

If a database query produces data which does not correspond to what the user requested, nor to any other legitimate reading of that query, then the system has failed. However, given a system where the user can customize, for example, the conceptual model of the database, all failures must be investigated for possible user errors in customization. The mechanical process of customization must also be investigated to ensure that the user is warned against making customization errors of this type.

4.3   Unexpected outputs

In cases where the user is presented first with a natural language interpretation to verify how the system is treating his input, there is a case for arguing that as long as this interpretation is clear and it corresponds to the command or query, then users do have a chance to change the input to get what they really want. The output should then be classified as unexpected, but not necessarily as a failure. For example, if a user of a database query system asked *'How big is the HMS Discovery?'*, expecting an answer relating to the weight of the ship, but received the interpretation *'What is the height of the ship HMS Discovery?'* and an answer in feet, then the system would have failed to answer the user's question, but would not have failed to give a reasonable answer and would not have mislead the user.[5] There is of course a point where such outputs deviate too much from what is expected and have to be seen as system failures. There is a discussion of such cases in Walker, Nelson and Stenton (1992) where they criticize a system which infers too much information based on predefined **world knowledge** and so gives unexpected answers.

4.4   Too much output

Natural language is ambiguous and so any natural language system must have a strategy for dealing with this. One method is to only ever give the user one possible reading for a particular input and hope that the user will find a way to rephrase the input to elicit a more appropriate response. Other systems offer users a choice of all possible readings and let the user pick the most suitable one. Using the latter approach can lead to situations where the user has to search through a large number of responses to find the one most closely resembling his intentions. Even if all readings are logically possible a user can become very frustrated, especially if the input seemed perfectly unambiguous. As with wrong and unexpected outputs, the particular application can affect ambiguity of outputs to a high degree for customizable systems. Finding the boundary of what levels of ambiguity are acceptable is not straightforward and must be judged from a customer perspective.

---

[5]   Given the appropriate information, the above example could be customized so that both interpretations would be presented to the user.

4.5　No output

Any natural language system can only hope to deal with some subset of a language and so there will be cases where the input from users cannot be handled. This implies the need for diagnostics of some kind. It is obviously better for the user to get an error message than to have the system give a badly-formed or wrong answer, but if the message does not help the user to understand how to rephrase and get the required answer, then the system has to some extent failed. For a fuller discussion of diagnostics see Zoeppritz (1992) in this volume.

4.6　Unexpected input

If the system is presented with unexpected input, that is if the input is ungrammatical, outside the domain of the customized application or even in the wrong language, then the output from the system should at least show the user:

- that the input was unacceptable

or, in the case of there being some unexpected analysis

- how that output was reached, so that the user is not mislead into thinking that the input was accepted as intended.

For example, if the user of a query system, using an application dealing with the scheduling of meetings asks *'What time is it?'*, then it would be reasonable for the system to interpret this as *'At what time is there a meeting named IT?'*. If the user is given a paraphrase corresponding to this interpretation, then the system is not misleading the user, if, however, the user is merely presented with the time of a meeting without explanation, then the system has mislead the user and has failed.

4.7　Error analysis in practice

We have come to the conclusion that working out what is acceptable and what is a failure is not at all straightforward. In an ideal world all these factors would be taken into account. Unfortunately, software development always has strict limits on time and resources. Often, compromises will be made in order to get the testing completed, for example, what constitutes an error may be strictly defined, If, however, too many of the factors listed above are ignored in that definition, then the quality of the resulting system will suffer. One way to get closer to the ideal situation is to have a complex system of flags assigned to test cases, rather than a binary system of OK/NOT OK. This certainly complicates the statistics of the test results, but does give a much clearer idea about how the system has performed. Problems must be prioritized, but it is also important that cases not classified as completely satisfactory are recorded somewhere so that improvements will eventually be made for these cases. Often the unclear cases are forgotten and only definite failures are handled.

5 The test team

The choice of the individuals who should test is as important as the choice of test tools. Testing is typically a low-prestige job and often the selection of testers does not take into account all the skills required. Natural language systems are often marketed as usable by more or less any speaker of that language, but this does not mean that just anyone can be an efficient tester. Testing is a vital and highly skilled task and should be viewed as such when selecting testers.

5.1   User viewpoints

According to most sources on software testing, the ideal test team is chosen from outside the development team. This is intended to ensure that the product is tested by individuals who are not biased by its internal structure. Pfleeger (1992) states very clearly the difference in viewpoints between the user and the programmer with respect to the functionality of the product. What the user sees as a failure may very well be justified by the programmer as a **design decision**. Having an independent test team for the later stages of testing is a way of approximating the user's viewpoint.

5.2   Testing as criticism

There is also often a reticence amongst programmers to admit to all problems and even more reticence to documenting all errors found in the code. Testing may wrongly be seen as a test of the programmer's skills, rather than just one more stage in the process. Not documenting all errors can bias final statistics, from which important decisions are made about future work. Such documentation is also often used to discover the cause of regression errors. If the documentation is incomplete, it will slow down the process of debugging.

5.3 Invalid errors

An independent test team should, in theory, produce a more complete documentation of errors. However, in practice the result of a completely independent team is a mixed blessing, since many of the failures reported will prove to be invalid after further investigation by the development team. This not only slows down development, but can cause friction between the teams. It is therefore advisable to have some kind of liaison between the two groups to filter all failures reported by the test team before they reach the developers. Invalid errors for a query system include:

User errors See the discussion above.

Duplication The failure is caused by an error already reported elsewhere.

Suggestion The input is outside of the coverage of the current version.

The last point raises the important question of determining the boundaries of coverage. This is a difficult and often controversial task. Defining the coverage of a system should be systematic, so the user will not have problems where one construction works fine, but another very similar one inexplicably fails. It is not a valid argument to say that a test case is outside of the coverage because that construction was simply overlooked by developers. It may be that there are time constraints that mean that a problem will be deferred until a later stage, since it is not considered critical to the functioning of the product. Labelling a problem as a suggestion should not, however, mean that the problem is forgotten. All problems must be addressed properly, though possibly for a later version of the system. The liaison should know enough about the internal structure of the system and the development team to be able to assign most problems to the appropriate developer. Whitten (1989) suggests that a good solution is to allow the testers access to a limited number of developers for consultation and to have regular follow-up meetings to ensure that the number of invalid problems reported decreases as the testing progresses.

The term **invalid** used in this context is not uncontroversial. Magdalena Zoeppritz pointed out to us (personal communication) that the term is likely to be offensive to testers, as it undermines the importance of their work. If testers are intimidated by long complicated explanations from developers, they may be discouraged from pointing out other (valid) errors. The term seems to put blame on the tester, rather than describing the planned action that developers will take. If a long explanation is needed to explain to a tester why the error is invalid, then some action will have to be taken to ensure that a user doesn't meet the same problem and require a similar explanation. Users will not accept long, technical answers to what they see as a failure of the system. An invalid error may be just the first sign of a deeper problem.

5.4   Economic factors

For small scale projects, having an independent test team may not even be an option due to lack of resources. For the reasons listed above, this is not an ideal situation, but given that the disadvantages of such a situation are taken into account, testing can be completed successfully by a subset of the development team. In this case, developers should be given a thorough education in the requirements of potential users, as this knowledge is vital for successful testing.

5.5   Language skills

For a natural language program, the testers of a system should be native speakers of that language or should at least approximate the linguistic abilities of the users. For example, if a Spanish product is produced in Spain, but will be used also in South America, at least some of the testing should be carried out by a native of some South American country. Likewise, a product developed by speakers of British English should also be tested by speakers of dialects from North America, Australia and anywhere else where the system will be marketed. The testers should also have a good knowledge of the output of the system, for example, the syntax of a command language or a database query language.

If the targeted user group is very specific, then there could be particular dialectal factors that should be taken into account. For example, experts in a particular type of data base query language may tend to use expressions and constructions based on that language. Questions may start with *select distinct* rather than the more expected *list* or *show me*. If, however, the targeted customers are very unlikely to have this type of knowledge, then a tester with this kind of bias can give a very wrong impression of the coverage of the system.[6]

---

[6]   See Zoeppritz (1985) for a discussion of the way users adapt their language when using a natural language system.

5.6 Size of the team

Other key factors are the size of the test team and the length of time that any individual will have to spend testing. Designing the test is a difficult, but potentially very interesting task. Performing the test is long, repetitive and tedious. If the test team is too small and individuals are forced to spend extended periods of time testing, then both the morale of the team and the quality of testing will suffer. Fatigue amongst testers can lead to mistakes and this can jeopardize the results of the test. Well-designed tools can certainly make the job easier, but the thorough testing will always be slow and repetitive and cannot be totally automated.

6 Test tools

Testing any software system requires many repetitions of the same type of action. Automatic tools are therefore essential to increase the efficiency of testing. For a natural language query system, what is required is a tool which will enter large numbers of questions and store the results for verification by the testers and for comparisons with subsequent test runs. The tool requirements for testing an operating system interface would be very similar. The first run of new test cases will require a tester to check each individual result. This task can be made easier by the use of user-friendly interfaces, but will always be time-consuming. If possible, this task should be divided amongst several testers to avoid fatigue errors. Well-designed comparison functions can greatly reduce the workload in subsequent test runs. For unit tests, it may be an advantage to run a subset of the modules, running and checking intermediate structures in order to pinpoint problem areas. Tools are also required for documenting and reporting errors for development. Robustness and performance must also be tested, both for the individual components and for the entire system.

6.1 Batch testing

A typical batch test tool allows testers to run large numbers of test cases, stores all relevant outputs and compares the outputs with any stored results. The tester should be able to flag test cases to verify their status. The tool should have a simple interface, which combines ease of use with portability to other environments (if necessary). Ideally tests should not interfere with development work and so should be run on either a different system or at a different time. Typically this means that tests are run overnight. Tools must be designed with recovery procedures so that the failure of one test case will not stop the rest of the test from being run. Similarly, a system failure should not lead to the loss of all results already produced.

In the case of a database query system there is a question of what results should be stored: the output query statements or the answers returned from running the queries against a database. All output statements should be run to verify that they do not have illegal syntax etc., but storing the answers rather than the query statements has disadvantages for test cases relying on contextual or deictic information. For example, some questions relating to dates give different answers according to the system time/date. For these cases, it is much easier to recognize the change in the query statements rather than by looking at the data returned from the database. Another drawback associated with storing answers is that each test run will take considerably longer to run and will use more resources. It is also possible for two very different query statements to give, by chance, the correct answer and so a significant change in the output structures could be missed if only answers were compared. Output structures saved should also include any error codes, so that these can be verified for potential problems. Storing intermediate structures can be very useful for debugging. It removes the need for developers to rerun the question for each failure and can sometimes help pinpoint the errors more easily. However, in practice we have found that storing more than the final structures causes problems because of the sheer size of the result files. Such files may be difficult to manipulate using some editors, and storing the results from several test runs requires a huge amount of storage. Since the results should be made public to aid debugging, it may not be feasible to store all structures for more than the most recent test run.

The design requirements for such a tool are actually very similar for the testing of other types of software. Griesmer (1990) describes a batch test tool for an expert system, which stores both the intermediate structures and the final answers produced by the system (for example, figures relating to tax and budget). In the case of this system it is possible to verify the answers using another type of tool designed for a similar task. This would also be possible for a natural language query system, since there are many other types of query systems. For the system described in Griesmer's paper, the intermediate structures are only checked if the answers have changed.

6.2   Intelligent comparisons

At first sight, the tool needs only the most simple comparison function to compare new results with stored and verified files. This proved to be inadequate since the changes to the system can produce many changes in the order of the output structures.The same phenomenon can occur if a system runs under different environments and a test is required to ensure that the same results are obtained. A simple comparison of results files will report that all structures have changed and will require the tester to check, by hand, which structures have only changed order. This can lead, in many cases, to other significant changes being missed.

If the tool stores all the intermediate structures, then a simple comparison of all structures will yield many changes that a user would never see. A change in, for example, a parse tree, which is not transmitted to the final structures, is of only limited interest and should not automatically be reported to testers. This information may be very important in the case of a failure to produce the correct final structures, but is irrelevant otherwise.

A more sophisticated comparison strategy is, therefore, needed to increase efficiency and accuracy. For the problem of structures produced in different orders in consecutive test runs, the strategy we chose was to match up as many structures as possible for a given test case and only report those structures which were missing from either the new or stored results. For example, if the new results produced structures: *A, B, C, D, E* and the stored result showed: *B, A, C, D, F, G*, then the following changes would be reported:

- New structures: *E*

- Missing stored structures: *F, G*

This leaves the tester with the task of deciding whether *E* is a changed version of either *F* or *G*, or if it is a totally new structure. The tester must also decide if the loss of one structure is an improvement or a regression.

Where all intermediate structures are stored, changes must be viewed with respect to any changes to the final structures. A possible strategy here is to look first at the end structures and only if there are changes there, check the differences in other structures. It is, however, difficult to devise a method of spotting only those changes that are relevant to the particular failure. This is crucially a task for developers to define, since detailed knowledge is needed of the intermediate processing stages. Reporting all changes may in fact slow down debugging as developers wade through masses of irrelevant data.

6.3   Updating

Once a test has been run, the testers must check the results; all results for a first run and all changed results for subsequent test runs. The ideal test tool will minimize the number of key strokes per test case and will have regular backups, so that data cannot be lost due to system failure. It is important that structures to be verified are displayed to users in the clearest way possible using **pretty printers** and taking into account the maximum size of a particular structure. It is very difficult for a tester to view a structure if it is much larger than a conventional screen and scrolling is needed in both directions. If structures are likely to be very large then the tester should have the option of printing them. This is time-consuming, but can give more accurate results. Verification and flagging should be completed, if possible, in the same stage and therefore on-line verification is usually preferable. For cases where there is more than one structure to be flagged for a particular test case, it would be useful to have some process of default flagging for all non-exceptional cases.

For test runs where results are being compared to older stored results, there should be automatic inheritance of flags for unchanged structures. The tester should then be presented with only those cases that have changed. At the point where a tester is checking a changed result, it should be possible to see both the new and old structures, with all flags. It has also proved very useful to see, at a glance, the number of interpretations for each structure. This helps the tester when faced with duplicate readings.

As mentioned above, the system of flagging structures should be more complex than a binary OK/NOT_OK system. For example, it is useful to mark all cases of duplicate structures for questions where there are several identical readings. It is also useful to note missing structures and to note which structures have problems. The tester should have the option to see, for a set of verified test cases, all test cases with a particular flag and boolean combinations of flags. Such information should, preferably, be delivered in the form of a report with statistics. For example, a tester may need a list of all cases not marked as OK or may want statistics for the percentage of questions marked as having problems with the paraphrases, excluding those cases of duplications.

6.4   Change predictions

Testers often have to deal with the phenomenon referred to by Whitten (1989) as **creeping functions**. This term refers to the common practice amongst development teams of continually adding functionality to a system, particularly during the formal test phase. This is desirable on the one hand since the user should get the benefit of as many new functions as possible, but if these changes are not documented in some type of **change control mechanism**,[7] then there can be severe problems for testers with tight schedules to test these new functions and check for regression errors.

If major changes come as a surprise to testers, then not only can testing be slowed down drastically, but also if there are simply not enough resources to deal with the re-verification of large numbers of test cases, the new function may simply have to be removed. Even if the change for each test case is minor and predictable, the impact on testing can be enormous. If, for example, certain words were suddenly to be capitalized in all or most natural language interpretations, then testers would have to check **all** changed structures. In checking such a large number of cases, there is a high risk that, due to fatigue, another problem in a changed structure may be missed and a structure wrongly marked as correct.

Developers must therefore ensure that testers know in advance of major expected changes and planning should take account of such changes in the assignment of test resources. Tools should also be adapted to deal with major predictable changes of this type. Ideally, it should be possible for developers to define the set of test cases that should be affected by the change so that the testers can verify that the change occurred correctly in all cases. If testers are only prompted by changes to test cases, there may be hidden problems such as test cases that didn't undergo the change, but which should have changed.[8]

6.5   Error reporting

Testers need to have some way of documenting problems and reporting them to the appropriate developer. We have already discussed the notion of having a liaison to deal with this and/or meetings, but there is also the need for tools to convey the information efficiently.

---

[7]   Typically a system for documenting proposed changes, having them inspected and approved before implementation.

[8]   We are grateful to Per Kristiansson for this last suggestion (personal communication).

Testers should not report each failure in the test, but instead should list the underlying problems where the problems are identifiable. Ideally, the tester should be presented with all instances of a particular type of failure at the same time. This may prove far too difficult to achieve automatically and so one possible solution would be to present the tester with an on-going log of unsolved problems to date during the verification phase. The tester could use this as an aid to classifying test results, so that duplicates of problems are not reported. The problem log should then have references to all test cases exhibiting a particular problem. This information is vital since for a problem to be declared as solved each of these test cases must be run successfully. Having the verification of test results and the problem reporting mechanism linked should make the testing more efficient.

The assignment of errors to particular modules and, ultimately, to individual programmers is often a complex task. Some failures are relatively easy to assign, for example any instance of an ill-formed structure is generally a problem in the final generation module for that structure. It could also point to defects elsewhere, but at least part of the problem is easy to track down. For most other types of failure, a good deal of knowledge about the structure of the system is necessary and a problem will often be passed from one developer to another until the error is finally located. The problem reporting system must therefore allow for the change of assignment of a problem and must ensure that the newly assigned developer is notified immediately and all relevant information sent to him or her, so that each developer does not repeat the same set of investigations. Failures can also be due to multiple errors, involving more than one developer. Similarly, error fixing can have a ripple effect, with one change forcing several other related changes.

7   Multilingual systems

Most software products are now adapted for use in many languages. For the majority of products this involves the translation of panels, manuals and any other written materials. There are sometimes problems with different character codes and maybe other environment changes to be considered, but by and large this is all that is needed to add another language.

With a natural language system the situation is completely different. In addition to all the tasks listed above, adding another language means rewriting large amounts of code using expert linguists. Unless the system has been specifically designed for multilingual use, it may also involve a complete redesign.

LanguageAccess is a multilingual product. The natural language engine has been specifically designed to be easily adaptable to other languages. It is composed of a series of modules most of which are used in all language versions. The parts which are language specific are: the grammars (analysis and generation), the base vocabulary and hierarchy, and the language model (created with a language specific version of the customization tool or CT). All other parts, including the semantic processing, SQL generation and the parser remain constant for all versions. The multilingual design of LanguageAccess is due in no small part to the influence of the User Specialty Languages system developed at the IBM Heidelberg Scientific Center. This prototype system had versions in several languages and most of these were adapted to the LanguageAccess system. There has also been language versions developed only in the LanguageAccess system. To date there have been at least seven different languages for which grammars have been developed. Of these languages the most advanced versions are for English and German, both of which have been released as products (English available since January 1991 and German since June 1991). Versions for French and Spanish are also well developed. Until now only Germanic and Romance languages have been developed, but feasibility studies for other language groups have been favorable.

7.1 Serial vs. parallel testing

As we have already emphasized, testing a natural language system is time-consuming enough for one language. Adding another language compounds the problem. So wouldn't it be easier to stick to one language at a time, test that one language thoroughly and then only have to test the specific language parts later? There are certainly advantages to this type of development. For example, all the interfaces, manuals and marketing materials can be finalized before sending them to the translators. This definitely makes life easier for the interface design team. However another argument often used in favor of serial, rather than parallel, development of language versions is that it is better to get the rest of the code stable before adding an extra complication, often using the work-load of the testers as evidence to support this. Our experience has lead us to disagree with this view. Whilst in the short run there will be savings in testing and elsewhere, serial development can lead to much more serious problems at later stages, which will have a far greater impact in terms of time and resources.

After working with several language versions in both serial and parallel modes, we have concluded that there are actually very few truly language independent changes that can be safely tested using only one language. The input to a natural language system is generally taken to be infinite in variety and there will be for each new language added structures which have never been produced before. Experience has shown that there are many faults that cannot be reproduced in all languages. This has lead to arguments of the type, "Why slow down the development of Z because of a problem that only appears in Y?" This argument may well be valid for the current version, but if the system is to be developed further and new functions added, it may well show that the language specific problem is in fact a design problem, which will affect other languages, too. Making the code as general as possible from the start, will, in the long run, save time-consuming redesigning of modules and unnecessary extra testing.

Stabilizing the linguistic code before several languages have been tested can lead to the situation where the design is so language-specific that it will be impossible to implement some functions for other languages. This will then force a rewrite of code, possibly even a total redesign, which will in turn require repeat testing for all languages. Given the tight schedules imposed on development projects, this can even lead to a function being unavailable for one or more languages. Once such a decision has been made it is often very difficult to get the resources at a later stage to do the necessary rewrites. Having different functionality for different languages has many drawbacks in terms of the interface requirements, for example, error messages. Testing a new function in several languages at once means that code will be tested with a larger number of syntactic variants and therefore more thoroughly than with only a single language.

We would not want to suggest that parallel testing is easier or faster than testing in serial. With our experience of multilingual testing, we know only too well how much resources and effort are required. Careful coordination is necessary to ensure that batch tests do not run simultaneously, draining the system resources and potentially slowing down all tests. Also, there must be enough testers with the appropriate language and other relevant skills available whenever necessary. This can be a big problem in a small development team where there are only one or two developers per language and limited resources available for an independent test team. Testing many languages at the same time is logistically difficult. The problem is still greater if the language versions tested are using different non-language specific code. It is much easier to test versions where the only differences are in the specific language components. Simultaneous testing of systems with different functionality levels leads to confusion as to which functions are implemented in which languages and so to problems of error reporting, not to mention the problems of developing and making fixes in more than one level of code.

Taking all these factors into account, we conclude that it is far more efficient, from both a testing and a development point of view, to test as many of the languages simultaneously as possible. This will require the assignment of more resources to testing, but in the long run, will prove to be much more efficient and will give better results in terms of reliability, which is, of course, the reason for testing in the first place.

7.2   Unit testing

The extent of multilingual testing can be reduced somewhat by the use of module-to-module unit testing. For example, a set of pre-generated intermediate semantic forms can be passed through later modules. Automatically generated forms could be used for this. This type of testing is very useful for testing the robustness of the code and to ensure that no forms with illegal syntax are produced. However the developer must be sure to include a wide variety of different forms as input and making predictions about what can be produced in a number of languages is a far from trivial task. It should also be noted that this type of testing is of a completely different nature to testing from questions and should be seen as a supplementary method and not as a substitute. It is essential to test functions from a natural language question and view the resulting structures in relation to that question and not just to the intermediate structures. The user of the product will not be remotely interested in anything except the question and the answer.

7.3   Code inspections

After the code has been module/unit tested, but before the formal testing has begun, most systems go through some form of code inspection or walkthrough.

This stage has been referred to as the **human testing**[9] phase since it involves the reading of code by humans rather than the running of code by machine. This is often fairly unpopular with developers since it is a long, arduous task, spanning several days for a large program and involving hours of looking at printouts of computer code. It is however, if performed well, one of the most productive stages for locating program errors. Myers (1979) states that in some cases as many as 80% of all errors found in testing a product are found in this part of the process. Typically a small number of fellow programmers sit down with the author of the code to be inspected, after each has spent several hours individually looking at the code. All errors found in the inspection are noted by a moderator.

So why is this stage particularly relevant to a natural language system? For a single language system the answer is probably: only as useful as for any other system. However, for a multilingual system, built in a modular way with language independent and language dependent modules, where some of the language dependent modules are specific to a particular language, the inspection phase gives a formal opportunity for all writers of a particular language specific module to compare ideas and locate potential errors. In a multilingual query system such as the one we have worked with (and in many translation systems) there are grammars for each language which share a particular formalism and also a common interface to the semantic components. We only have experience of relatively similar languages, that is Germanic and Romance languages, but given that a system can be successfully adapted to a very different language, there is nothing to suggest that the same principles would not also hold true. The constraint of having a common interface makes all grammars of a system share many common features and in our experience it has proved very useful to exploit the similarities as much as possible. For example, given a grammar formalism where the order of rules is unimportant to the running of the program, it is very useful to maintain the order of grammatical structures handled in much the same order for all grammars. This makes the task of inspection far more efficient. Often in the earlier stages of the development process, it is useful for the writer of one grammar to check how a structure is handled for another language. Careful ordering can save time in these cases and will often help to avoid the situation where each grammar writer

---

[9]   See Myers (1979).

'reinvents the wheel' for common constructions. Simultaneous inspections of grammars for two languages have also proved to be very useful, given a common organization within the grammars and a high degree of similarity between the two languages.

The usual criteria for the involvement of a programmer in an inspection is that the programmer should be skilled in the programming language and should know something about the product. For the inspection of grammars this translates to: should be skilled in the grammar formalism. Knowing the natural language under discussion is obviously an advantage, but has not proved to be at all essential. Given a well documented grammar, it should be possible to review most of that grammar and locate a high percentage of errors with only a very scanty knowledge of the language. Some modules would certainly require more specific knowledge about the natural language to give useful comments, for example, the base dictionary components and the morphology definition component for user-defined vocabulary.

The intended use for the formal inspection is to locate errors rather than to come up with solutions to these errors. However, in practice code inspections provide programmers with a useful opportunity to discuss design issues. Myers (1979) also quotes an example of this type of deviation from the intended purpose of code inspections.

Since less formal inspections and walkthroughs prove to be so useful for specific language code, it would seem reasonable to include such a phase in the development process for multilingual systems, to be held at an earlier stage, before the design has been finalized and the program unit tested.

8 Conclusions

Above all, testing is difficult. There is no easy way to test a product successfully. The quality of a system depends crucially on the choice of testers, test cases and tools. There are no short cuts. Since development teams have been testing software for many years, their knowledge of testing should not be ignored. Natural language systems have some special requirements, but more traditional testing methods can be applied with good results. If natural language systems are to succeed in the market place, there must be empirical evidence of their reliability, performance, and usability. Such evidence can only come from high quality testing.

9 Acknowledgements

We would like to thank many people for their help and support during the writing of this paper. We are particularly grateful to Magdalena Zoeppritz, without whom this paper would not have been written. We would also like to thank the entire natural language processing department at IBM Nordic Laboratories for their words of encouragement and for being such a great team to work with.

10 References

[1] Aron, J. D. (1983): The program development process: Part II The programming team, Addison-Wesley Publishing Company.

[2] Griesmer, James H. (1990): "Validating and testing the FAME expert system" in The proceedings of the ITL conference on expert systems Yorktown Heights, N.Y.

[3] Horgan, Joseph R. and Aditya P. Mathur (1992): "Assessing testing tools in research and education" in IEEE Software May 1992.

[4] Malhotra, Ashok (1975): "Knowledge-based English language systems for management support: an analysis of requirements" in ITCAI-4.

[5] Musa, John D. (1989): "Tools for measuring software reliability" in IEEE Spectrum February 1989.

[6] Myers, Glenford J. (1979): The Art of Software Testing John Wiley and Sons Ltd, N.Y.

[7] Pfleeger, Shari Lawrence (1992): "Measuring software reliability" in IEEE Spectrum August 1992.

[8] Sanamrad, Mohammad A. and Ivan Bretan (1992): "IBM SAA LanguageAccess: A large-scale commercial product implemented in PROLOG" in Proceedings of the 1st international conference on practical applications of PROLOG, London, April 1992.

[9] Walker, Marilyn A., Andrew L. Nelson, Phil Stenton (1992): "A case study of natural language customisation: The practical effects of world knowledge" in The proceedings of the 15th International Conference on Computational Linguistics, Nantes, August 1992.

[10] Whitten, Neal (1989): Managing software development projects: formula for success, John Wiley and Sons Ltd, N.Y.

[11] Zoeppritz, Magdalena (1985): "Computer Talk?" Heidelberg Scientific Center TN 85.05

[12] Zoeppritz, Magdalena (1986): "Investigating human factors in natural language data base query" in Jacob L. Mey. Language and discourse: Test and protest. A Festschrift for Petr Sgall John Benjamins publishing Co., Amsterdam/Philadelphia.

[13] Zoeppritz, Magdalena (1992): "Software ergonomics of natural language systems" (this volume)

Computertalk and Multimodality

Jürgen Krause
IZ-Sozialwissenschaften
Lennéstraße 30
D-53113 Bonn
krause@iz-bonn.gesis.d400.de

1   Problem space

Besides the continuing activities of the last 25 years in constructing natural language interfaces by simulating the human natural language capacities on machines and the upcoming of its "natural" alternative at the beginning of the eighties, the graphical interfaces, a lot of recent research activities in computer science, artificial intelligence and information science center around multimodal computer systems. Mixed modalities (e.g. text and graphics) are one of the most favored ideas nowadays in spite of the fact that the area is still in its formative stage (cf. Maybury 1991 as an overview).

With respect to human computer interaction (HCI) mixed modality means more than the integration of the two basic "natural" modes of HCI: the graphical and the natural language mode. Nevertheless handling the question of how to mingle both basic types can be seen as a prototypical problem space for the theoretical and practical questions connected with this field.

At first glance multimodality does not appear to be too much connected with the first term in the title "computertalk". Roughly speaking, the term addresses the differences of language use between human communication and natural language HCI (and the question whether such a difference exists at all). To show empirically the existence of computertalk in HCI is the main theme of my paper. The connection to graphical HCI and to multimodality lies in the way of interpreting these differences (cf. section 4). Beyond the starting concept of a language register computertalk I will argue that language in HCI is used metaphorically - not in 1:1-analogy to human communication - in a similar sense as the metataphorical use of the desktop metaphor in graphical HCI. Therefore, discussing the basic theoretical concepts of graphical HCI is helpful for a better understanding of natural languge HCI. Additionally this interpretation will give us a common point of departure for developing a theoretical model of HCI which includes both "natural" modes, graphics and natural language, - a result, that could be considered as only subsidiary from the viewpoint of computational linguists only interested in 'pure' natural language algorithms but which is indeed decisive for multimodal HCI.

2 The basic "natural" modes: natural language and graphics

2.1 Natural language

The "naturalness" of natural language interaction is established by the fact that users are already familiar with this (almost literally self-explanatory) mode of communication. The underlying thesis - widespread in computational linguistics and artificial intelligence - says: There is a 1:1 analogy between the user's behaviour in HCI and human communication (cf. e.g. Grishman 1986, Kannngießer 1989). The need to learn new ways of interacting, prevalent in command-oriented systems (such as SQL) is eliminated by the user's already existing and well-trained skills in human communication.

Following this thesis two problematic aspects arise:

a) No natural language interface covers the whole range of human communication. Thus the question is, whether the partial solution designed for an application area meets the requirements of the actual retrieval situation. In the worst case the advantages of taking over knowledge from human communication get lost, if the handling of the implemented subset of language requires learning and recalling efforts comparable with those of formal language alternatives.

b) It is not yet clear whether users in a special application domain use the same natural language utterances and show the same behaviour in HCI as in human communication. If differences ("computer talk") exist, they have to be determined empirically and must be considered in the design of the natural language input component.

Problem a) leads to the subset discussion, which shall not be discussed further here (see Krause 1982). The subset problem can be seen (although not necessarily) as a practical difficulty in technologically transfering natural language algorithms to software products, resulting from the fact that today and in the near future this transfer has to be done before the linguistic knowledge and the edp development has been sufficiently advanced to enable the adequate simulation of human communication. But these arguments are not valid for problem b). It questions whether the model of 1:1 analogy between human communication and natural language HCI is theoretically adequate and practically useful.

The thesis of an almost 1:1-analogy is a very simple design thesis; it is also very practical because the only thing the software developers have to do in order to construct a natural language interface - "pure" or mingled - is to copy the existing knowledge of linguistics, the knowlegde of human communication.

But what will happen if the thesis of the 1:1-analogy is not correct, if users do not behave in HCI as they do in human communication, e.g. use different syntactic constructions, different words or different dialog strategies?

In this case natural language algorithms imitating human communication will be constructed, which would only work in an interface, where the users look upon machines as another kind of human beings. But if the users of natural language interfaces do not think so, if differences exist, we have to explore them in order to come to feasible solutions.

2.2 Graphical/direct manipulative interaction

This second basic form of "natural" HCI can mainly be characterized - apart from the use of icons, pull-down menus, and windows - by two features:

a) Underlying metaphors

A central but simple insight, provided by cognitive science, justifies the use of metaphors: new phenomena (new knowledge) are easier to learn and remember, if ties to knowledge that is already present exist. For the domain of wordprocessing and office communication the physical office environment represents such a tie. Therefore the screen is designed as a desktop and the functionality of wordprocessors is realized in analogy to the familiar typewriter. For functions that go beyond the desktop or type-writer (e.g. the clipboard concept) the whole office is taken as a metaphor for electronic objects (icons representing bookcases, folders, paperbaskets etc.). Thus the work of the user is simplified; he can draw conclusions in analogy to the familiar office environment.

b) Direct manipulation and mouse
The term "direct manipulation" was coined by Shneiderman , who also gives the classic example of this principle of HCI in Shneiderman 1983. He characterizes the difference between graphical interaction and conventional (command-oriented) systems by a comparison with car-driving as a prototypical example of the application of direct manipulation. Instead of using function keys to determine the desired direction (*"right"*, *"left"*, specification of angles) or giving natural language input (*"turn the steering wheel 30 degrees to the left"* ...) we turn the steering wheel itself. We get an instant feedback of the changes caused by the action and can perform appropriate corrections. Instead of verbalizing we act immediately.
In the same way the user operates by means of visual objects and the mouse as the pointing device on his electronic "desk" or in his "office". He is explicitly encouraged to think in physical (instead of electronical) terms and real actions.

It would be an illusion to think that these techniques and theories will lead by themselves to systems that can be used without mistakes and require no training at all. The use of the mouse has to be trained in order to perform the desired movements precisely. Neither is the technique of clicking (single-click, double-click, holding down the button in pull-down menus) self-explaining nor can learning be dispensed with.

Beyond that the theory of metaphors implies - different to the postulated 1:1 analogy of natural language usage discussed above - that there will always be deviations, that is, violations of the metaphor. For instance, the electronic desk in office communication does not really correspond with the real desktop in every respect and the analogy with the typewriter is incomplete at best. The electronic world is equal to the real office-world only insofar as there are analogies with a lot of details that help the user in becoming familiar with the functions of the software.

2.3   Conclusion

The two primary modalities for the design of "natural" user interfaces are based on different theoretical assumptions which are similar in the sense of being based on the simple but central insight provided by cognitive science: new phenomena are easier to learn and remember, if ties to knowledge that is already present exist. But there are important differences which can be expressed by the terms "1:1 analogy" versus "metaphorical use".

As natural language HCI will not work, if differences between human communication and natural language HCI exist which are neglected by following the thesis of 1:1 analogy, the next question to be answered is, whether there are differences between human communication and natural language HCI at all. As the answer to this question is important for all kinds of "pure" or mingled interfaces, and as the majority of computational linguists believed and still believe in this analogy and build up their systems on this basis - neglecting possible differences - we have tried to prove the existence of such differences empirically, that is, the existence of something like computertalk, in the DICOS project of the LIR (funded by the German Ministery of Research).

3   The computertalk experiments

Dicos tried to find answers to the following three problem areas:

a) Differences in formulating natural language utterances between human communication and HCI

b) The consequences of constructing restricted natural language HCI software

c) The necessity to find a common point-of-view for looking at both natural interaction modes.

3.1   Test design and experimental factors

DICOS consisted of a hidden-operator test, meaning that there are two rooms, one for the test subject, one for the examinator and his technical support like videomonitor, tape recorder, videomixer and so on. In the subject's room the test is supervised and recorded by two videocameras and a microphone: the examinator simulates the tested machines of each application.

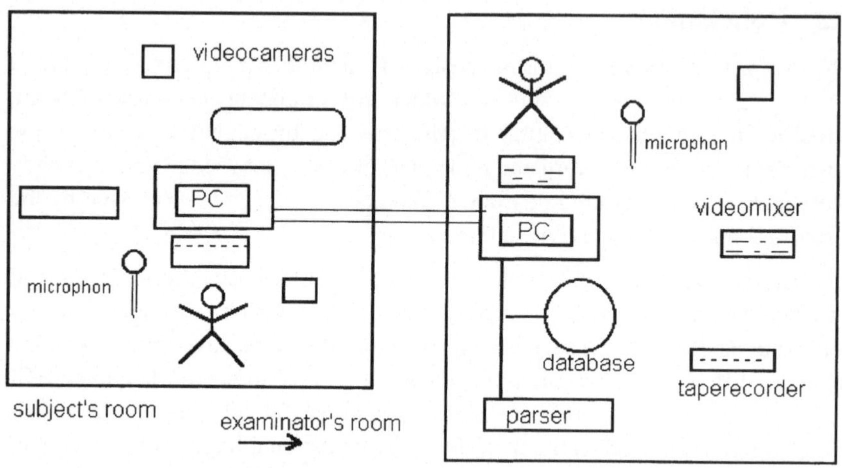

DICOS includes two different application domains: information seeking dialogues within a railway information system and within a library environment. Two hidden-operator experiments were conducted, in each of which four information systems with different capabilities representing the capabilities of present day and future information systems were simulated.

System 1:

The user is told that he is communicating with a railway or library employee, who does the lookup in a database and answers his questions through a computer as a special input/output channel. The system has no restrictions in language utterances or cooperativity. The user gets an echo describing his information need.

System 2:

System 2 behaves exactly like System 1, but the user is told that he is communicating with a computer understanding natural language without restrictions. The user can ask whatever he wants. The only difference between System 1 and System 2 is the mental concept of the system in the head of the user.

<table>
<tr><td>

System 3:

The cooperativity of System 3 is restricted. The user has to formulate his information need according to given restrictions (no vague or modal expressions, literally interpretation of time phrases etc.). If the user does neglect the restrictions, his utterances are rejected and he is told by the system which mistake has occurred.

System 4:

System 4 introduces additional restrictions. Above that it behaves like System 3, but performs no error analysis.

</td></tr>
</table>

The experimental design involved two experimental factors:

a) the input-medium (voice input or keyboard input) and

b) the system variation (System 1, System 2, System 3, System 4) as discussed above.

This led to an experimental design with eight cells and five subjects each:

	System 1	System 2	System 3	System 4
voice	5	5	5	5
keyboard	5	5	5	5

For each experiment (railway and library) 40 test persons had to perform eight tasks, which took them about two hours. The chosen tasks represented different scenarios as shown in the example below:

"Your infirm grandmother wants to travel to Hamburg. Find a travelling possibility as convenient as possible for her".

Statistical experiments set up hypotheses so that comparisons concerning the factor levels can be carried out by so called hypothesis tests. Within these tests the null hypothesis saying that there is no difference between the factor levels is compared with another hypothesis. The null hypothesis can be rejected, if a difference will be shown within a specified probability of making a wrong decision. In DICOS statistical hypotheses were formulated contrasting human communication with HCI (S1 vs. S2, S2, S3), human communication with a non-restricted computer system (S1 vs. S2), a restricted communication situation with a non-restricted situation (S1, S2 vs. S3, S4), and voice input with keyboard input (S1, S2, S3, S4 voice vs. S1, S2, S3, S4 keyboard).

3.2   Empirically based starting hypothesis

In statistical experiments it does not matter, where the hypotheses come from. One can test the silliest hypotheses. The only effect will be wasted time. In contrast to that, the statistical experiments of the LIR-tests in DICOS are only one element in an overall evaluation plan, where 'weak' interpretative methods are used to gain plausible starting hypotheses, tested statistically at the next level.

From a methodological point of view there were two main phases of research activities. Phase 1 contained real world field studies with the experimental natural language interface USL, developped at the IBM scientific center in Heidelberg (cf. Lehmann 1978, Jarke/Krause/Vassiliou 1986 and Krause 1990 for detailed description).

In the first phase it was tried to find in an interpretative manner empirical indications for differences between human communication and HCI in protocols of real world applications with USL between 1978 - 1987. The protocols analysed contain about twelve thousand questions of more than 100 persons.

3.3   Language register "computertalk"

A second task of phase 1 was to find a first adequate model to describe the empirically found differences between human communication and natural language HCI. The method chosen to interpret the USL-protocols and later the DICOS-tests was that of a language register like foreigner talk or baby talk.

What does that mean? Registers like foreigner talk or baby talk are mainly seen as a special kind of simplification of "normal talk". A set of different parameters of the language-in-use situation is correlated with a set of structural features (like missing morphological endings or articles). The basic idea is, that not only do people talk differently in different situations, but that they do it predictably: specific types of situations have definable properties, wich in turn have determinate consequences on language. Register theory now wants to make explicit the links between the situational features and the linguistic features (cf. Ferguson 1985, Halliday 1973, Paris/Bateman 1991).

In our context the two most important modifying processes of register theory are simplification and clarification (cf. Ferguson/deBose 1977).

Simplification means that there is a reduction on the surface structure. People omit parts of the lexicon or grammar rules to ease understanding for the dialog partner - e. g. when his language competence is restricted.

Clarification processes also intend to make language understanding easier -- but not by a reduction of surface structure. On the contrary clarification adds redundancy to the message. It increases the substance of a sentence by supplying material that is normally omitted.

Simplification and clarification can plausibly be correlated to the special language-in-use situation of HCI; where the addressee is a non-human computer considered by the user as having some kind of special competence but perhaps also a restricted natural language competence, as well as other incompetences, associated with its being human.

We found a lot of such features already in the USL-studies, mainly looking at

> *"several instances of deviant or odd formulations that looked as if they were intended to be particularly suitable for use with a computer as the partner in communication."(Zoeppritz 1985:1)*

Some examples of the USL-studies are:

a)　Simplification

> *List the mst code and the last name all the alumni in New York City. (missing proposition)*

> *List the last name and the first name of the alumni whose last name with S-. (missing verb)*

> *List the address, donation in 1979, donation in 1980, donation in 1981 for the alumni with zipcode like '20 %'. (missing coordination)*

> *List the last name, first name, addresses, donation of y. (missing coordination)*

b)　Clarification

> *... gave 'more than 0' in 1981*

> *... gave 'donations' in 1980*

> *... who live in 'the city' los angeles*

... with activity 'for year' 82

... members that are not donors 'of donations' in 1982

(The expressions in single quotation marks are unnecessary; the quotation marks are added by the author)

Nearly one third of the evaluated questions in the protocols of the USL-studies were classified by two native speakers as aberrations of normal human communication. Therefore one aim of DICOS was to verify these hints and some others from literature by statistically valid hypothesis experiments.

3.4 Summary of the DICOS results

The full details of the DICOS tests can be found in the articles of Krause/Hitzenberger 1991. Here only some results important for the thesis named in section 1 are summarized. There were mainly five differences between S1-S4 which I want to report here:

a) There is a successive tendency from S1 to S4 to restrict the variation of sentence patterns. Their complexity and amount are highest in S1, the human dialogue situation. They are less in S2, the unrestricted HCI, and lead partly to question form patterns, where only slots (mainly the specifications of place and time) are filled differently.

b) Indirect interrogation including imperative, declarative and optative clauses is highest in S1 and lowest in S4, in contrast to the direct interrogative clauses which are low in human communication and become high in man-machine dialogues S2 - S4, a tendency which is augmented by the restrictions in S3 and S4.

c) The interpersonal features of human dialogue diminish in the HCI-dialogues of S2-S4, such as the explicit initiation of a dialogue with salutation and polite closure, modal particles (*may be, if possible*) and polite subjunctive formulation of questions (*Could you please tell me*).

d) Features - especially in spoken language - which are interpreted as dialogue disturbances or errors are reduced in S2 - S4 in contrast to the human communication in S1 (e.g. different kinds of ellipsis like abandoning a construction or subsequent correction, additions and appositions). The users seem to feel more obliged to express themselves correctly when the addressee is a (stupid) computer than when speaking with humans.

The four feature groups are examples for the difference of human communication (S1) versus HCI (S2-S4), whereby in most cases restrictions seem to intensify the computer talk effect. But there are also indications of special features of restricted systems (S1 - S2 vs. S3 - S4):

e) The elliptical constructions are clearly higher in the unrestricted systems S1 and S2. This seems to be an effect of the tendency to surpass the restrictions of S3 and S4. As the user knows, the systems are restricted, he is (also unintentionally) afraid to leave the frontiers of the language capabilities. He is also afraid to speak about extraordinary topics. E.g. in S1 and S2 a topic is asking for the valid conditions of certain reduced tariffs. In S3 and S4 users avoid such topics.

Dicos did not succeed in achieving statistical significance for the hypothesis tests with respect to all interpretatively gained features presented here and in Krause/Hitzenberger 1991.

Statistically significant results could be achieved in two cases, which is surprisingly much given the actual state of the art:

a) The reduction of the features that signal "computertalk" showed significant results (95% level) concerning voice and keyboard input. There is a significant difference between natural language HCI and human communication in the usage of these features.

b) The analysis of variance also showed significant results with regard to the partner oriented dialogue signals. People use more of them when talking to a human partner than when they talk to a computer system which behaves exactly like the human informant.

3.5 Conclusion

Summarizing the observations of DICOS we can claim that the existence of differences between human communication and HCI has been proven. Language restrictions of the natural language HCI system and the user's mental model of his dialogue partner (machine or human) influence the choice of language formulations. There is no 1:1 analogy. There are special rules and regularities that cannot be captured by looking at linguistic grammars or by observing human dialogues.

4   Metaphorical use of language in HCI

The conclusion drawn above contains only one part of the findings of DICOS. Besides these indications there were also some examples of HCI-discourse, which led to the thesis that computertalk means more than some aberrations describable within a linguistic model of language use.

4.1   An example

Here we can only give an example (cf. Krause/Hitzenberger 1991, Krause 1990 for further material), first published in Krause 1982 as an isolated anomaly (without further theoretical interpretation).

No.   Question

14   *Wieviele Schüler haben welche Englischnote in Quarta?*
             (How many students have which grade in English in seventh?)

Note	Anzahl
(Grade	Number)
1	22
2	77
3	144
4	112
5	42
6	2

....

29   *Welche Deutschnote in Quarta hat wieviele Schüler?*
             (Which grade in German in seventh has how many students?)

> U32410W: SPALTENNAME: VON NICHT ENTHALTEN IN TAB: SCHÜLER
>
> (U32410W COLUMN NAME: VON NOT FOUND IN TABLE: STUDENT)

30   *Wieviele Schüler haben welche Deutschnote in Quarta?*
        (How many students have which grade in German in seventh?)

The question number 29 is not acceptable in German. This did not, however, bother the user. He chose this form because the answer to question number 14 showed Note ('grade') in the first column and Anzahl ('number of the students') in the second.

The user wanted to make it easier for the computer to answer questions of the type number 14. Therefore, in number 29 he adapted his query to the (correctly) assumed internal sequencing.

It is important that this destruction of an acceptable language structure, which proved to be successful in HCI, took place without being required. There was no error situation and no strange looking result.

To interpret this and similar examples adequately we have to admit that the user leaves our language system as a whole. He does not vary it (or it cannot be explained as such) by taking something away, adding a small new rule or simplifying regularities of general language.

He choses a vantage point outside of his own language competence and appeals to the idea that a computer cannot talk: 'it's only a trick of the designers'. From this external position he uses language parts as movable scenery, primarily subject to the general logic of computer actions, and not to language competence, (slightly) adapted to the needs of the language-in-use situations.

If one accepts this interpretation, another or an additional concept has to be found to handle these situations: It seems that natural language is used here in the same metaphorical sense as the desktop metaphor. The user builds up a mental model of his natural language HCI, which is based on the picture of human communication in natural language but which is not in one-to-one correspondence.

In the context of this section the most important property of the metaphorical concept is that metaphors can break. Metaphors do not work in all cases, but very often nevertheless.

4.2 Conclusion metaphorical usage

Besides the prove of differences between human communication and natural language HCI in the sense of a language register computertalk, there are also strong hints for the metaphorical use of natural language, similar to that of the graphical mode when using the desktop metaphor. The user builds up a mental model of his natural language HCI, which is based on the picture of human communication in natural language but which is not in one-to-one correspondence.

Accepting this interpretation we have obtained the common point-of-view we searched for to interpret natural language HCI and the graphical interaction techniques. Both can be based on the concept of metaphorical use.

5 References

[1] Ferguson, C, A. (1985): Editors's Note. Special Language Registers: Discourse Processes 8, no. 4 (special issue). 392-394.

[2] Ferguson, C. A., DeBose, C. E. (1977): Simplified Registers, Broken Language, and Pidginization. In: Valdman, A. (ed.). Pidgin and Creole Linguistics. Bloomington/London. 99-125.

[3] Brishman, R. (1986): Computational Linguistics. An Introduction. Cambridge et al.

[4] Halliday, M. A. K. (1967): Explorations in the Functions of Language. London.

[5] Jarke, M., Krause, J. (1985a): New empirical results of user studies with a domain-independent natural language query system. In: Bibel, W., Pettkoff, B. (eds.): Artificial Intelligence Methodology Systems Applications. North-Holland. 153-159.

[6] Kanngießer, S. (1989): Korrespondenzen zwischen KI und Linguistik. In: Luck, K. v. (Hrsg.): Künstliche Intelligenz. 7. Frühjahrsschule, KIFS-89. Berlin et al. 270-282.

[7] Krause, J. (1982): Mensch-Maschine-Interaktion in natürlicher Sprache. Tübingen.

[8] Krause, J. (1990): The Concepts of Sublanguage and Language Register in natural Language Processing. In: Schmitz, U., Schütz, R., Kunz, A. (eds.): Linguistic Approaches to Artificial Intelligence. Frankfurt a. M. et al. 129 - 158.

[9] Krause, J. (1991): Empirical Indications about the Existence of a Language Register 'Computer Talk'. In: Bammesberger, A., Kirschner, T. (eds.): Language and Civilzation. Regensburg. 755-778.

[10] Krause, J., Hitzenberger, L. (Hrsg.) (1991): Computer Talk. Hildesheim/Zürich/New York.

[11] Lehmann, H. (1978): Interpretation of Natural Language in an Information System. IBM-Journal of research and development no. 5. 560-572.

[12] Maybury, M. (ed.) (1991): Intelligent Multimedia Interfaces. Workshop Notes from the Ninth National Conference on Artificial Inteeligence (AAAI-91. Anaheim, California. July, 15, 1991.

[13] Paris, C. L., Bateman, J. A. (1990): User Modeling and Register Theory: A congruence of concerns. Paper Information Sciences Institute, Marina del Rey, California.

[14] Shneidermann, B. (1983): Direct Manipulation: A Step Beyond Programming Languages. IEEE 8. 57-69.

[15] Zoeppritz, M. (1985): Computertalk? IBM Heidelberg Scientific Center TN 85.05.

What a Grammar of Spoken Dialogues has to Deal With

Hans-Ulrich Block and Stefanie Schachtl
Siemens AG
Corporate Research and Development
Otto-Hahn-Ring 6, D-81730 München
{block,Stefanie.Schachtl}@zfe.siemens.de

1   Introduction

The promising results speech recognition achieved in the recent years confront computational linguistics with a new task. The attractivity of speech understanding systems in the market justifies the effort to cope with spoken language computationally. This article is intended to sharpen the awareness of the difficulties that lurk behind the knowledge transfer from text to speech understanding. The comparison of the grammars of spoken and written language, especially the surplus of regular constructions in spoken language are the subject of several linguistic works, reviewed here to set up a sort of syntactic agenda for speech understanding systems. Nothing is said however about the various methods that will help to fulfill this agenda.

2   Methodological Preliminaries

The goal of this article is to give an overview of the syntactic properties that are specific for spontaneous spoken dialogues. In order not to mingle ourselves in universal philological discussions with little practical use we decided in favour of a data driven approach with our task. First we investigated the DICOS data for spoken language[1] with respect to specific syntactic behaviour and grouped those specifics together. Wherever possible we refered to the literature available. Eventually, we also investigated syntactic constructions that were cited in the literature but were not documented by the data. Still the choice of the constructions discussed appears pretty much haphazard but we hope to have

---

[1] [DIC:1], [DIC:2], [DIC:3], [DIC:4]

mentioned at least all of those constructions that are relevant in the imminent task of understanding spontaneous speech dialogues.

A normative definition of 'what is spoken language' was not available as a basis for our investigation. Accordingly, we decided in favour of a practical approach: we concentrated only on those constructions that are common in spontaneous conversation which however text understanding systems of today are incapable to handle.

The data stem from various data collecting efforts with the subject of train schedules, cf. ASL, FACID, DICOS. DICOS assembles dialogues that were recorded from individuals who were confronted with four different simulated NL advisory systems both in written and spoken conversation. The first of these systems was explained to the user as being truly human, the second was said to be a computer that allegedly understood everything that was said, the third and the fourth system appeared considerably reduced in their language understanding capacities. Those reductions concerned first of all that the utterance of only one single sentence at a time was allowed, ellipses were prohibited, metaphors etc. were not understood. Subordinate clauses could be treated only on the first level of embedding. The fourth system mirrored the actual state the SPICOS system in 1989 was in, which included the before mentioned reductions and added to them a general verdict of subordinates, relative clauses excluded. Modalverbs, inexact quantifiers and other vague expressions were not understood.

3   Discrepancies in Quantity

Already [Wun:84] focuses specific syntactic behaviour of conversational language. He lists under the superscript of typical appearances of conversational language: 'initiations of dialogue', 'economy efforts (ellipsis, aposiopese)' and 'lavish behaviour (supplements, reiterations and accumulations)' . Already at the turning of our century [Beh:27] accentuated the view of two distinct systems of spoken and written language. Among the studies of more recent date the work of [Wei:75] and [Hoh:75] focuses syntactic aspects of spoken language. Hoehne-Leska compares in her statistic investigation spoken and written monologues of standard German with respect to the features of length of the sentence, sentence structure, length of dependent constituents, length of words, number of constituents, number of appearance of certain constructions, constituents and subordinates, number of substituteable attributive parts, and content of information (see [Hoh:75]:29).

The study was based on 50 spoken and 50 written texts. In her summary Hoehne-Leska mentions most significant differences between spoken and written language.

Length of sentence   The average length of a sentence in written texts was 19-20 words, the average length of a sentence in spoken texts amounted to 13-14 words. The difference is more significant with respect to the distibution graph. 40% of the data in the spoken texts comprise a sentence length of 5-10 words. The maximum frequency (25% of the data) lies within 6-8 words. In written texts sentences of a length of 7-19 words constitute the main part of the data, the maximum frequency lies within 1-,2-,3- and 15-word sentences.

Sentence structure   The share of simple sentences within written texts Hoehne-Leska gives as 56%, within spoken texts it reaches 65%.

Form and function of constituents   Whereas discrepancies in the function of constituents are only marginally discernable in written vs. spoken texts, the form of constituents reveals strong differences between both sorts. According to [Hoh:75] the following constructions were remarkably preferred in written texts: infinitve constructions (in the ratio of 1:2.11), participle constructions (in the ration of 1 : 2.45), the 'Stirnform' ('V1 clause' , see 4.1) (in the ratio of 1:3.12). In spoken texts the 'Kernform' ('V2 clause') is more used (in the ratio 2.13:1), (see [Hoh:75]:99). Concerning the dependent subordinate clauses a significant discrepancy can be noted with respect to the conjunctions *wenn, weil* and *da*. In spoken texts *wenn* appears about twice as much as in written texts, *weil* is preferred in the ratio of 2.41:1 by speakers. On the other hand *da* is the conjunction that is most often used in written texts (in the ratio of 2.75:1)

Attributive constructions   Discrepancies with respect to that feature are very significant. On the average there appear 67 attributive parts in constituents within the spoken language, in the written language we find a mean value of 226 (that is the ratio of 1:3.36). In spoken language attributes mostly belong to adjectives whereas in written language attributes are more likely to be represented by nouns and word groups.

Although the work of Hoehne-Leska accentuates the differences between written and spoken language with respect to the quantitative distribution of certain constructions quite strongly it is still not clear whether or not there exist constructions that are specific for either of the text sorts. This is critisized by ([Wei:75]:12) who deplores the absence of the description of specific spoken language constructions.

The work of Weiss itself concentrates on quantitative characterizations of language variations as well but compared to Hoehne-Leska it has the advantage that Weiss first extracts the specific features of spoken language with which he describes different sorts of texts in the second part of his work.

Weiss starts out with two assumptions. First he explains how a speaker navigates between the two polarities of standard and substandard language. The speaker commands over a set of language levels that are in between those two poles. Here standard language means the best approximation of his level of standard language towards the normative standard (see [Wei:75]:14). Second he is aware that the situation and subject of a conversation exact strong influences on the syntactic behaviour of the produced speech. The areas where syntax is sensitive to this influence Weiss denotes as the

- number of simple and complex sentences

- form and number of subclauses

- form and number of syntactically incomplete sentences

- use of shortened and elliptical setting

- appearance of change of construction of various kinds

- frequency of reiterations, anticipations and supplements

(see [Wei:75]:1).

Relatively consistent with the results of Hoehne-Leska Weiss states the extremely high percentage of simple sentences in spoken language. On the average 76.8% of the sentences in the texts he analysed were simple sentences. Among the remaining complex clauses only 20% comprised more than one subclause. That shows that the overall percentage of complex clauses with more than one subclause with 5% is very small.

Weiss' statistic results fortify that speakers in "stressy" conditions approach their respective level of standard language whereas in normal conversation they tend to use their "normal", substandard level. This tendency is also mirrored in the data of the DICOS project ([DIC:1], [DIC:2], [DIC:3], [DIC:4]). The language performance level in [DIC:2] most significantly approaches standard level when compared to [DIC:1] (human vs. computer as conversational partner). The consciousness (or at least the imagination) of talking to a computer in itself brings about this 'stressy' behaviour in the speaker[2].

We would like to assume for the following discussion the criteria of segmentation Weiss introduces: He determines 'Äußerung' ('utterance') as the largest exactly documentable unit of a spoken text. He defines 'utterance' with [Ric:66] as

> jede lautsprachliche Produktion eines Sprechers in einer einheitlichen äußeren kommunikativen Situation (d.h. beispielsweise bei einem Dialog bis zum Einsetzen des Partners) ohne Rücksicht auf Binnenstruktur und Länge[3] (see [Wei:75:20]).

While the unit of 'utterance' can still be reliably defined, segmentations underneath that level are hard to be discerned. Most interesting in this context is Weiss' experience that segmentation into minor parts according to criteria of tone level and boundary, i.e. prosodic criteria did not lead towards a comparable syntactic segmentation (see [Wei:75:17]). He argues that the optionality of prosodic marking is responsible for that (see [Wei:75:25]).

As a syntactic unit Weiss only hesitantly accepts 'sentence' . A 'sentence' counts as a syntactic unit also in those cases where an 'utterance' or part of an 'utterance' constitutes a syntactic unit and commands over an own semantic content. Syntactic unit and semantic content are crucial even then when those two conditions are fulfilled only with regard of the context (see [Wei:75:23]). Opposite the 'sentences' we find the so called 'contact words' such as *ja, aber*, that induce 'utterances' very often.

---

[2] see [Kra:91]), who documents differencies in the language behaviour of test persons depending on the fact whether they communicated with a human or a computer.

[3] every speechlike production of a speaker in an externally uniform communicative situation (i.e. for example in a dialogue to the point where the other dialogue partner utters something himself) regardless the internal structure and length

4   Constructions that are Typical for Spoken Language

4.1   Basic Assumptions Concerning the Grammar

In order to clarify the following discussion we shortly introduce a few basic assumptions on German syntax.

The referring category for many syntactic constructions is the clause. We differentiate clauses according to the position where the finite verb is situated: In German we find clauses with the verb in second position (a), in the first position (b) and in final position (c):

(1)   a.   Karl *fährt* nach Hamburg.

         b.   *Fährt* Karl nach Hamburg?

         c.   ... daß Karl nach Hamburg *fährt*.

Those clause types we call 'V2 clause' (a), 'V1 clause' (b) und 'Ve clause' (c).

Complex verbs in 'V2' and 'V1 clauses' constitute a so-called 'Satzklammer' ('brace construction')[4] bracketing the finite verb in second or first position and the infinite parts of the verbal complex in clause final position.

(2)   Karl *ist* mit dem Zug nach Hamburg *gefahren*.

According to the position of the verbal elements three different 'Stellungsfelder' ('positional fields') of the clause are discerned. The 'Vorfeld' ('topic field') comprises the area in front of the finite verb in a 'V2 clause' . The 'Mittelfeld' ('middle field') comprises the area internal of the 'brace construction', cf. the area between the finite verb and the infinite verbal parts in a 'V2 clause' . The 'Nachfeld' ('post field') comprises the area behind the verbal elements in final position up to where extraposition starts.

topical	finite verb	middle field	infinite form postfield
Karl	*ist*	*mit dem Zug nach Hamburg*	*gefahren*
	Ist	*Karl mit dem Zug nach Hamburg*	*gefahren*

---

[4]   Specific technical terms will be introduced by their standard expression from German grammar. The English translation always refers to exactly this meaning of the expression.

Keeping this basic structure of a German clause in mind we are now going to investigate amplifications of this structure that are typical for spoken language.

4.2 Amplifications to the Left

4.2.1 Clause Introducing Elements

One of the most frequent phenomena in spontaneous dialogues is the introduction of a sentence via a single word or tag such as *und, also, aber, na, beziehungsweise, und zwar, das heißt, sagen wir mal, wie gesagt* etc. or via a filling element such as *äh, ähm* etc..

The following is a list of possible word classes these introductory elements may stem from.

Coordinative conjunctions   The coordinative conjunctions *und, oder, aber* etc. occupy a fixed position in front of the 'topical field' in spoken as well as in written texts. Their usual duty is to combine expressions of equal value (words, constituents or clauses). In (3) *aber* is positioned in front of the 'topical field' of the second clausal conjunct which is occupied by *Maria*:

(3)   Peter ist nach Hamburg gefahren, *aber* Maria war nicht da.

In substandard use sentences can also be introduced by a coordinative conjunction without there being an adequate first conjunct in reach. Sequences as given in (4) are not seldom documented:

(4)   Ben.: Nennen Sie mir die Abfahrtszeiten der Züge nach Hamburg!

Sys.: <tells dates>

Ben.: *und* wann fährt ein Zug von Hamburg zurück?

Neither the first sentence uttered by the user (which was an imperative) nor the answer of the system (which consists in a list of dates) represent a possible first conjunct for the second sentence uttered by the user. Nevertheless he uses *und* in front of the 'topical field' of the second sentence uttered.

However, the data confirm the assumption that it is probably not possible to initiate a whole new dialogue starting with *und*.

Subordinative conjunctions   Here the substandard use of *weil* is to be mentioned especially. *Weil* tends to be used like the coordinative conjunction *denn*.

(5)   Ich möchte eine Fahrkarte kaufen, *weil* ich muß nach Hamburg fahren.

Conjunctive adverbs   Adverbs such as *nun, also, bloß, nur* represent an important group of introductory elements that tend to deviate from their normal behaviour in written language. In spoken language those adverbs are often treated like real conjunctions. They appear in front of instead inside the 'topical field' of a clause.

(6)   *bloß* i hob' keine Ahnung welche ... (D-1 vp213-4)

Since the function of adverbs and particles is dependent on their position in the clause conjunctive adverbs may change their function when used in this way: *Nun, ich brauche eine Tasse Kaffee.* does not mean *Nun brauche ich eine Tasse Kaffee.*

Answering particles   The use of the answering particle *ja* in front of the 'topical field' is twofold: First *ja* may constitute an answer to a previous question. In this case *ja* represents an independent prosodic group and is separated by a prosodic boundary from the following sentence. It therefore represents a sentence by itself. This behaviour does not deviate from written texts.

(7)   Sys.: Sind Sie unter sechsundzwanzig?

          Ben.: *Ja.* Bekomme ich dann eine Ermäßigung?

The second and in spoken language by far the most frequent use of *ja* consists in its function as an introductory element. There it appears prosodically integrated in the following sentence. It has no propositional value.

(8)   *ja* dann paßt' s ja (D-1 vp213-3)

Filling elements   tend to show up at the beginning of a sentence:

(9)   *äh* . können Sie mir bitte die Verbindungen . von . /äh Regensburg nach
          Frankfurt . am Abend . sagen ... (D-1 vp216-2)

          ähm muß ich da eigentlich dann auf dem direktesten Weg nach Kiel
          fahren ... (D-1 vp216-6)

Filling elements are determined by the idiolect of the respective speaker. The DICOS data p.e. allow an obvious distinction between *äh*-users and non-*äh*-users.

Tags   Several tags are used in introductory functions, cf. *wie gesagt* and *sag'n wir mal* in the following examples.

(10)   *sag' n wir mal* die nächsten beiden Züge würden mi' a no interessieren (D-1 vp213-1)

ja also . *wie gesagt* / . ich möcht'Nach Nürnberg fahr' n ... (D-1 vp213-1)

Combinations of introductory elements   The possibility of combining introductory elements is quite restricted. [Wei:75] lists combinations of conjunctions and adverbs such as *und dann, und jetzt, ..., aber auch, weil jetzt* etc. However, it is not made clear whether within those combinations the adverb keeps its substandard position in front of the 'topical field' or whether it appears inside the 'topical field' itself, thereby retaining its adverbial function. The latter case reveals itself the more probable because of the inacceptance of constructions like: *und dann ich brauche noch eine Fahrkarte* even in substandard contexts.

We also like to maintain in accordance with [Wei:75] that combinations of two conjunctions or two adverbs with one another are not or only marginally possible.

In contrast to that combinations of answering particles with a conjunction or adverb respectively (cf. *ja und, ja also*) are abundant. The answering particle regularly occupies the first position in this combination.

Tags can be combined with combinations out of conjunction and adverbial elements, cf. (10).

4.2.2 Prominence Structures on the Left of the Clause

Spoken language is strongly characterized by constructions as exemplified in (11) and (12). Here constituents appear outside their regular position in front of the 'topical field' or behind the 'brace construction' .

(11)   *des Supersparpreisticket* **des** haut net hin

(12)   **des** haut net hin / ähm *des Supersparpreisticket* (D-1 vp213-5)

[Alt:81] gives a detailed description of those constructions. Following his terminology we call forthwith 'Herausstellungstrukturen' ('prominence structures') expressions that confirm to the following critera (see [Alt:81]:46f):

- 'prominence structures' are no complete clauses in the formal syntactical sense. They only gain their own semanto-pragmatic function via the clause they are adjoined to either before or behind.

- They are connected with the adjoining clause but they are never completely integrated into it.

- They don' t fulfill any function in the adjoining clause, therefore they can' t be obligatory members of the adjoining clause.

- They don' t fill 'positional fields' .

- Usually they are positioned at the left or right margin of a clause.

Altmann differentiates between 'prominence structures' on the left and 'prominence structures' on the right of the sentence. Subsumed under 'prominence structures' to the left are 'Linksversetzung' ('left dislocation'), 'freies Thema' ('hanging topic'), 'vokativische NP' ('vocative nominal group') and 'Wiederholung' ('reiteration').

left dislocation   Constructions with 'left dislocation' are characterized by a prepositional or nominal group appearing in front of a normal 'V2 clause'. This clause includes a pronominal element that is coindexed with the 'left dislocation' expression and appears in the 'topical field' of the clause. The 'left dislocation' expression and its pronominal are coadjacent.

(13)   *Die Brigitte*, die kann ich schon gar nicht leiden. (see [Alt:81]:48)

There is no prosodic boundary between 'left dislocation' expression and 'V2 clause'. It is possible to combine several 'left dislocation' expressions, in case that iteratively stronger references follow each other (see [Alt:81]:129).

(14)   *Du und ich, wir beide,* wir schaffen es schon. (see [Alt:81]:129)

Apparently the 'left dislocation' expression can be combined freely with introductory elements:

(15)   *Und die Brigitte,* die kann ich schon gar nicht leiden

The 'V2 clause' the 'left dislocation' is adjoined to does not allow any introductory element.

hanging topic   'Left dislocation' and 'hanging topic' differ from each other mainly trough the existence of an prosodic boundary between 'hanging topic' and the following 'V2 clause'. Accordingly expressions that represent a 'hanging topic' possess their own sentencial prosodic marking. Furthermore the pronominal resumption in the 'V2 clause' is not necessary. The existence of a nominal group that may be associated to the 'hanging topic' suffices. Tags may introduce a 'hanging topic' .

(16)   *Apropos Pferde,* hast du Peters neue Stallungen schon gesehen? (see [Alt:81]:49)

In combining with each other 'hanging topic' always has to precede 'left dislocation' . A clause that follows a 'hanging topic' may contain an introductory element, however, the possibilities are quite restricted.

(17)   *Die Brigitte, also* die kann ich schon gar nicht leiden.

vocative nominal group   Related to the 'hanging topic' is the 'vocative nominal group', which contains as a special marker a pronoun of the first or second person:

(18)   *Ich Träumer!* Jetzt habe ich doch tatsächlich den Zug verpaßt! (see [Alt:81]:51)

reiteration   'reiteration' constructions as exemplified in (18) are also quite similar to 'left dislocation' and 'hanging topic' . They mostly function as contact inducers, introduction of discourse or intensivation (see [Alt:81]:52).

(19)   *ich,* ich lasse mir das nich gefallen. (see [Alt:81]:52)

It is important to note that this type of reiteration as a form of 'prominence structure' has nothing to do with the reiteration of parts of the clause in form of a 'new onset', which will be discussed in 5.1.

4.3   Amplifications to the Right

4.3.1. Ending Words and Tags

A certain amount of words or tags can be used to signal the end of a sentence. Those, however, are much less used than the introductory elements which were discussed in 4.2.1.

The interjection *ei* is often found prosodically integrated at the end of a sentence, its use is very substandard:

(20)   Das ist aber teuer *ei*

        Bringst du mal das Bier *ei*

Conjunctions and adverbs   like *oder, wohl, wohl nicht (wonich)* appear quite often at the end of a sentence in certain dialects of German. The DICOS data show the likewise use of *oder*.

(21)   und es ist irrelevant dabei . in welche Richtung man sich dabei bewegt . *oder*? (D-1 vp219-4)

bitte in its several varieties appears quite often at the end of an imperative or question clause

(22)   Wann fährt denn der Zug ab, *bitte*?

4.3.2 Prominence Structures on the Right of a Clause

The amplification to the right stretching over the end of the 'brace construction' is a phenomenon that is quite often met with in spontaneous speech. This fact is documented in the work of [Eng:74), who offers a comparison of data of written and spoken language. First Engel denotes the percentage of 'brace construction' constructions in those texts. From those he extracts the amount of constructions where the 'brace construction' is extended. Both values, however, do not reveal a significant difference between speech and written text. But if the existing 'brace construction' extensions are compared closely it can be observed that the extension via a simple lexical element or a nominal constituent are about five times as frequent in spoken language as in written texts (see [Eng:74]:221). Significantly strong are the findings with respect to prepositional complements following the 'brace construction' . Those appeared about 17 times more often in the spoken texts. 'Brace construction' extensions via subclauses on the other hand did not reveal any relevant discrepancies and are in fact responsible for the levelling in the summarizing data of right dislocation.

Also in the topic of 'prominence structures' on the right of the clause the work of [Alt:81] is fundamental. He distinguishes between 'Rechtsversetzung' ('right dislocation'), 'vocative nominal group', 'reiteration', 'apposition', 'parenthesis', 'extraposition', 'Ausklammerung' ('exbraciation') and 'Nachtrag' ('supplement'), of which only 'right dislocation' , 'vocative nominal group' , 'reiteration' and 'supplement' count as real 'prominence structures' . 'Vocative nominal group' (*Jetzt hab' ich doch den Zug verpaßt, ich Trottel*) and 'reiteration' (*Du hältst die Klappe, du*) act according to their left side equivalents, see the discussion in 4.2.2.

We shall now concentrate on the frequently used constructions 'right dislocation', 'extraposition', 'exbraciation' and 'supplement' .

Building on the analysis presented by Altmann [Aue:91a] sets up a new classification of those constructions that has its foundations not so deep in syntax but in the viewpoint of conversation as a sequence of parts that are shared by the speakers in a turn taking system (see [Aue:91a]:3) The basic assumption is that the change of the right to speak up is unproblematic only where syntactic endings are present. This leads to those phenomena of expansion that do not refer to syntactic constituents but to syntactic ending. Spoken substandard language is characterized by expansions that do not have to be clause-equivalent. Referring to a cross classification of syntactic and prosodic markings [Aue:91a] sets up a typology of not clause-equivalent expansions to the right.

	not clause-equivalent expansions to the right					
	regressive			progressive		
	syntagmatic	paradigmatic		parenthetical	conjunctional	others
		pronominal	others			
integrated	*ex-braciation*	*right dislocation*			*conjunct*	
not integrated	*supple-ment*	*right dislocation*	*reparition*	*apposition*	*conjunct*	*sequence*

The parameters of the above cited table are defined as follows:

- **regressive vs. progressive** A progressive expansion leads the preceding structure on, a regressive expansion modifies its preceding structure.

- **syntagmatic vs. paradigmatic** A syntagmatic expansion is integrated in the preceding structure, a paradigmatic expansion replaces parts of its preceding structures.

- **parenthetical vs. conjunctive vs. others** The expansion is not morphologically or syntactically marked. It is not marked by a conjuncive element or any other element.

- **prosodically integrated vs. not prosodically integrated** The preceding structure represents together with the expansion a single rhythmic phrase vs. two phrases: a) There exists a prosodic boundary between the preceding structure and the expansion. b) The preceding structure does vs. does not possess a final tone contour by itself. c) The expansion has vs. has no phrasal accent itself.

'Supplement' and 'exbraciation' differ according to Auer only in their prosodic behaviour, whereas Altmann counts 'supplement' among the 'prominence structures' but not the 'exbraciation' .

exbraciation   The elements that constitute an 'exbraciation' are positioned in the 'post field', i.e. behind the 'brace construction' and in front of 'extraposition'. They therefore belong to the clause itself. 'exbraciation' constituents may functionally depend on the main clause as adverbials or prepositional objects, some dialects even permit the 'exbraciation' of accusative objects. The 'exbraciation' need not be signalled by tags or similar expressions, there is no prosodic boundary between the main clause and the 'exbraciation' . Sentence stress lies on the element in the 'exbraciation' . A few typical examples follow, ' @' denotes the position that would be occupied by the constituent in the ' middle field' .

(23)   und wie schauts @ aus *a bisserl später* .. (D-1 vp213-1) jetzt bräucht i an Zug der mi @ nach Dortmund bringt *von Regensburg ab* (D-1 vp213-6)

kann man dabei auch die Fahrt @ @ unterbrechen *für mehrere Ta / mehrere Stunden* . in Köln zum Beispiel? (D-1 vp219-6)

extraposition   'extraposition' is no specific spoken language construction. We nevertheless will discuss it shortly because it helps in clarifying the difference between 'exbraciation' and 'supplement' . 'Extraposition' is solely defined as the position where subclauses in the main clause may appear, i.e. the position behind the 'post field' . There is only one 'extraposition' per clause available (see [Alt:81]:137).

(24)   Ich habe gesagt, daß ich eine Fahrkarte nach Dortmund brauche.

Können Sie mir einen Zug nennen, mit dem ich am Samstag nach Hamburg fahren kann?

No tags or similar elements are allowed within the 'extraposition', there is no prosodic boundary between main- and subclause (see [Alt:81]:66)

right dislocation   The 'right dislocation' mirrors the 'left dislocation' in clause final position. A clause-equivalent expression is isolated at the end of a clause.

The preceding sentence contains a pronominal refering to this expression. The boundary between clause and 'right dislocation' is prosodically marked. A typical tag that signals 'right dislocation' is: *ich meine*. In contrast to this adverbs, particles or similar elements are not acceptable (see [Alt:81]:54)

(25)   *des* haut net hin / *ähm des Supersparpreisticket* (D-1 vp213-5)

    die ist also unter keinen Umständen möglich . *so a' Verlängerung*

supplement   'supplement' constructions originate via reduction. They are heavily elliptic, clause equivalent and syntactically closely connected with their preceding sentence (see [Alt:81]:70). In a sense 'supplement' mirrors the 'hanging topic' in sentence final position.

Typical for the 'supplement' is the tag *und zwar*, adverbs may appear also as supplement introducing elements. The 'supplement' position is situated behind 'right dislocation' and 'extraposition', there exists a prosodic boundary between 'supplement' and the preceding sentence. Very often 'supplement' expressions possess the function of carrying on and are similar in function to expressions in the 'exbraciation'

(26)   ... und ohne Umsteigen einfach @ durchbraust . zumindest mal /meinetwegen bis Hamburg ... (D-1 vp213-7)

    könnten sie mir eine Verbindung @ . von Etterzhausen . nach Hindelang . äh durchgeben . und zwar für Samstag morgen . möglichst früh . ... (D-1 vp216-4)

A second important function of the 'supplement' expressions is to modify previously uttered parts of speech. In the following examples the 'supplement' expression determines a certain phrase in the preceding sentence more closely.

(27)   die ist also unter keinen Umständen möglich . *so a' Verlängerung . also a net in Ausnahmefällen*

    und zwar möchte ich nach Hindelang fahren . *von Etterzhausen nach Hindelang* ... (D-1 vp219-1)

Unlike the self correction via the agrammatical 'new onset' (see 5.1.) the 'supplement' manifests a perfectly grammatical construction in German.

4.4   Discontinuities

German is characterized by a multitude of discontinuous phenomena in written as well as in spoken text.

4.4.1 Long Distance Dependencies

Examples like (28) and (29) are fairly well discussed in computational linguistics and basically are known how to be dealt with:

(28)   a.   *Wann,* sagten Sie, *fährt der Zug?*

       b.   Welcher Zug, sagten Sie, fährt am Donnerstag nach Hamburg?

(29)   so ab sechs / hat sie g' sagt / wird sie wahrscheinlich losfahren (D-1 vp213-3)

4.4.2 *Was ... für*

Substandard forms of language differ gradually from the standard according to which long distance constructions are still acceptable. Among these counts the separation of *was für* in German. Instead of the standard construction (30) we often find long distance constructions as in (31)

(30)   *was für* Züge gibt' s denn da?

(31)   *was* gibt' s denn da / . *für* Zügerl? (D-1 vp213 3)

4.4.3 Preposition Stranding

Not standard but very common in spoken language is the sort of discontinuities called preposition stranding. In those constructions we find the pronominal and its preposition split from each other. The pronominal migrates to the left of the clause and the preposition stays behind (strands). So instead of (32) we find at least in northern parts of Germany very often (33).

(32)   a.   *Wohin* fährt dieser Zug?

       b.   *Was* für einen Zug kann ich denn *damit* benutzen?

(33)   a.   *Wo* fährt dieser Zug *hin?*

       b.   Was kann ich denn *da* für einen Zug *mit* benutzen?

4.4.4 Discontinuous Attributes

The split appearing of nominal constituents that form one syntactic unit represents an unsolved problem for all sorts of grammatical descriptions be it linguistically or computationally. In some cases that are also known from written texts nominal attributes migrate to the left of the clause and leave their head noun behind (b) and vice versa (c). This migration is also to a certain extent dependent of the verb of the main clause (d-e):

(34)   a. Eine auffallende Vorliebe für schnelle Wagen zeigte er.

        b. Für schnelle Wagen zeigte er eine auffallende Vorliebe.

        c. Was für eine Vorliebe zeigte er für schnelle Wagen?

        d. Von einer auffallenden Vorliebe für schnelle Wagen berichtete er.

        e. *Für schnelle Wagen berichtete er von einer auffallenden Vorliebe.

In quantifying constructions this sort of discontinuity is standard:

(35)   a. Von diesen Tomaten nehme ich fünf Stück.

        b. Wieviel Stück nehmen Sie von diesen Tomaten?

Another discontinuity that involves attributes and is very common in spoken language is exemplified by constructions like: *dort die Burg, vor ihm der Sarg, nebenan das Abteil.* Those constructions were investigated according to their quantitative properties by [Ulv:74]. He shows in his empiric databased analysis that those constructions although more common in spoken language also appear in written texts. The data stem from a collection of prose literature, a collection of broadcast texts and substandard data form various dialects. Ulvestadt constates that the semantic role of the attribute influences strongly the possibility of fronting it. 95% of the evidences in the written and broadcasting texts involved a local adverbial (see [Ulv:74]:279). In the substandard texts a likewise predominance of local adverbials is not given.

4.4.5 Discontinuities with Determiner and/or Adjective

Another frequently used construction of discontinuities concerns quantifying expressions in nominal groups. They sometimes get 'stranded' while the rest of the nominal constituent is found in the topic position of the clause.

(36)   **Züge nach Frankfurt** fahren heute *keine* mehr.

This construction poses further problems. The quantifier has not necessarily the same grammatical form as would be required in an adjacent position in the nominal group. (37) shows these oppositions:

(37)   a. Er hat *kein Bier* getrunken.

b. *Bier* hat er *keines* getrunken.

c. **Bier* hat er *kein* getrunken

a. **Er hat *keines Bier* getrunken.

4.5 Ellipsis

Ellipsis is one of the most important phenomena to be dealt with in the computing of dialogues. We will first discuss the regular ellipsis. Here the speaker leaves out certain elements in his utterance and expects the hearer to reconstruct those out of the given context (see [Kle:85:1]). Specific for spoken language dialogues are topic ellipsis, adjacency ellipsis and reduction in typical use.

4.5.1 Topic Ellipsis

A clearly definable group of ellipses is represented by constructions where the categorial filling of the 'topical field' is dropped. In standard language there is a strong functional discrepancy between 'V1' and 'V2 clauses'. For declaratives 'V2' always has to be used. However, declarations may also be made with verbs in front position in spoken language. Those constructions may be differentiated with [Aue:91b] in 'proper' and 'improper fronted verb clauses'. In 'proper fronted verb clauses' all necessary constituents are present and follow the finite verb. In 'improper fronted verb clauses' a single obligatory constituent is missing. It is assumed that this constitutent underwent ellipsis in the topic position.

Proper fronted verb clauses are characterized by the deletion of an expletive *es* or substandard *da, dann* in topic position. A typical context for *es* deletion are jokes, cf. (*Kommt ein Mann in einen Bäckerladen und bestellt ein Brötchen. Sagt der Bäcker ...*).

Improper fronted verb clauses are considerably more frequent than the proper ones. The obligatory constituent that is deleted in the 'topical field' has to be a pronoun. Its function can be either: subject (a), clause substitute *das* (b), accusative object (c) or dative object (d).

(38) a. Hast du Peter besucht? Nee, bin nicht da gewesen.

b. Ist Peter da gewesen? Ist möglich.

c. Hast du Peter getroffen? Nee, hab' ich nicht getroffen.

d. Und dem Chef? Hab ich' s schon gesagt.

The topic ellipsis can combine with the before mentioned discontinuities of prepositional proforms cf. 4.4.3. (39) shows this gradual development from standard (a) via prepositon stranding (b) to the cryptic but very natural sounding (c):

(39)   a.   Was hälst du von einer Fahrt in die Berge? Dagegen hab' ich nichts.

       b.   Was hälst du von einer Fahrt in die Berge? Da hab' ich nichts gegen.

       c.   Was hälst du von einer Fahrt in die Berge? Hab' ich nichts gegen.

4.5.2 Reduction in Answering a Wh-question

Dialogues abundantly show reductions of the following type:

(40)   Sys.: Möchten Sie den Fahrpreis für die einfache Fahrt oder für die Hin- und Rückfahrt wissen?

       Ben.: für die einfache Fahrt (D-1 vp216-1)

Both, question and answer are seen as an adjacency pair (see [Kle:85]:4), where in the second part all redundant information is dropped.

This can yield quite complex construcions:[5]

(41)   A: Wer hat was getrunken?

       B: Arnim ein Bock

4.6. Indepenent Settings

In some cases the main clause is replaced by a phrase that does not contain a finite verb. Although those clause equivalents are understood perfectly by the hearers no regular ellipsis strategy is discernible since the appropriate context is missing. [Wei:75:36] calls this instance a 'selbststaendige Setzung' ('independent setting'). A description of those constructions is found in [Gro:68]. He exemplifies how whole dialogues can be carried on without a single finite verb:

(42)   Kaufmann: Guten Morgen.

       Kunde: Morgen.

       Kaufmann: Bitte schön?

       Kunde: Einmal Lord.

---

[5]   see [Kle:85]:4

Kaufmann: Mit oder ohne Filter?

Kunde: Ohne, und ein kleines Dixan; so - ach, noch ein Dreipfundbrot.

Kaufmann: Von heute?

Kunde: Nein, von gestern, und hundert Gramm Wurst.

Kaufmann: Geschnitten oder am Stück?

Kunde: Gleich so.

Kaufmann: Noch etwas, bitte schön?

Kunde: Nein, danke.

Kaufmann: Fünf Mark zwanzig, alles zusammen.

Kunde: So. (bezahlt, indem er 6 DM hingibt)

Kaufmann: 6 Mark. Und 80 Pfennig zurück; vielen Dank, auf Wiedersehen.

Kunde: Guten Tag. (see [Gro:68]:65)

4.7 Parenthesis

Parenthesis usually denotes an insertion into a clause that is syntactically independent from this clause. Which forms of parenthesis are allowed is still under discussion. The possibilities range from from small interjections [Bay:73] to whole sentences, cf. the following examples:

(43)   ... und am Sonntag müßt' i / *na gut* / vor Mitternacht noch zu Hause sein

       ... (D-1 vp213-1)

(44)   ... ich möcht' nach Nürnberg fahr' n .. um die Museen zu besichtigen .wie da steht und . naja . also . so nach' m letzten Museumsbesuch *ich würd' sag' n* . so um nachmittag um fünfe wird des zumachen . da würd' i gern wissen . wann da ein Zug heimgeht ... (D-1 vp213-1)

([Alt:81]:63) also states that whenever a subclause appears in the 'middle field' in German it has to be parenthetical.

(45)   ... i würd heut' gern *.weil i grad Zeit hab'* mir die Fahrkarte kaufen (D-1 vp213-8)

Besides the categorial filling of the parenthesis itself the positions where parentheses may appear in a clause are of interest as well.

[Alt:81] notes the possibilities as:

- between 'topical field' and finite verb,

- following the finite verb,

- on constituent boundaries in the ' middle field' ,

- never between 'brace construction' and 'post field' ,

- very likely at the end of a sentence.

The DICOS data, however, exemplify series of other possibilities: we found parentheses between 'left dislocation' and 'topical field' (45), between main clause and 'extraposition' (46) and even between a contracted preposition and noun (47).

(46)   ähm heißt das . *wenn ich dann zurückfahr'* . da kann also ich nur freitags zurückfahren ... (D-1 vp216-6)

(47)   und also amm / . *ja sagen mer mal* Freitag abend und Samstag früh (D-1 vp216-4)

5   Ungrammatical phenomena

Up to now we discussed syntactic constructions that lie within the reach of a descriptive grammar of German. The remaining section of this article is devoted to phenomena usually named "illformed". The frequency of these phenomena in spoken language, however, cannot be ignored. Hearers obviously tolerate a certain type of bugs a speaker makes, especially those where he corrects himself within the same utterance. Those self corrections present a difficult problem for understanding systems.

The following typology of syntactic mistakes follows approximately [Wei:75]. A first division of ingrammatical input can be made between 'Neuansatz' ('new onset'), 'Satzbruch' ('sentence break') and 'Kongruenzfehler' (agreement mismatch). 'New onset' determines a set of mistakes where a phrase is interrupted by the speaker himself and the phrase is started anew with perhaps the same lexical material.

(48)   und was kostet die ganze / *was kost' a Rückfahrkartn* in diesem Fall? (D-1 vp219-1)

A break in a sentence means its construction gets changed during the utterance. The following example shows how the constructions of the sentences *Ich möchte eine Fahrkarte (kaufen)* and *Eine Fahrkarte möchte ich kaufen* are blended.

(49)   ich möchte eine Fahrkarte möchte ich kaufen.

Agreement mismatches finally include all sorts of morphological mismatches.

5.1   New Onset

Expressions in spoken language are characterized by a multitude of utterances where the speaker does not finish a sentence or a phrase properly but starts instead anew. [Wei:75] gives the percentage of new onsets in this sense with 28%.

New onsets can be clearly divided according to whether a new sentence, a new phrase or a new word is started. New onsets tend to be introduced by some sort of tag or filling element:

(50)   äh kann ich da eigentlich dann . äh / *dann* noch wei / also *wenn ich
        jetzt in Kiel ankomme* und hab' jetzt Lust zum / was weiß ich . *einen
        Abstecher nach Hamburg jetzt zu machen* . kann ich da mit diesem / *gilt
        dies dann auch oder muß ich dann eine* (D-1 vp216-6)

5.1.1   New Onset of a Sentence

A new onset of a sentence is characterized by the fact that a chain of words $w_i \ldots w_j$ that may represent a beginning of a sentence but does not constitute a whole sentence in itself, is followed by a new sentence. If the first chain $w_i \ldots w_j$ is properly included in the sentence, i.e. if the new sentence starts with $w_i \ldots w_{j+1} \ldots$ we shall call this a ' repeating new onset' of a sentence, cf. (48). If only parts of the old start are included, i.e. $w_i \ldots w_k$ $(k < j)$ this is called a 'proper new onset' of a sentence. The interrupted chain of words we shall call false start.

(51)   a. äh kann man diese einfache Fahrkarte . *kann man die
           Gültigkeitsdauer da verlängern* (D-1 vp216-3)

Sometimes a 'proper new onset' has no common parts with the false start:

(52)   und i . bräucht' no / *möcht' i da wissen . wie lang ist die Fahrkarte*
        gültig? (D-1 vp219-1)

5.1.2 New Onset of Phrases and Words

The division between complete repetition and partly repetition of 'new onsets' is also useful with phrases and words. The most common form of a 'new onset' on the phrasal level is the substitution of an unfinished word, this is called 'Wortabbruch' ('word break')

(53) a. wann kommt denn der Zug .. in dem i in der Früh' ei / *eingestieg' n* bin in Nürnberg / äh in Dingsda in Köln überhaupt an? (D-1 vp213-6)

A different strategy is the replacement of correct words, which can be induced by a tag:

(54) a. i bräucht also do auf' m Zettl eine Auskunft über / mit allen möglichen Fahrterminen ... (D-1 vp219-2)

b. i bräucht also do auf' m Zettl eine Auskunft über / das heißt *mit* allen möglichen Fahrterminen ...

Similarily to the 'word break' behaves the 'phrasal break'. Here a chain of final words in an utterance are replaced by a 'new onset'. The very last word in the false start is a 'word break':

(55) ... und zwar über eine Fahr / *eine Verbindung* nach Oldenburg ...   (D-1 vp219-5)

'Repeating new onset' of words or phrases is a special instance of 'phrasal new onset', where the words of the false start and the 'phrasal new onset' are identical:

(56) was ist die / *äh die* frühere Verbindung vor achtzehn Uhr achtzehn? (D-1 vp216-2)

5.2 Anakoluth

The blending of sentence constructions that does not fall under the various instances of new onset is often referred to with the term 'anakoluth' . We divide between two instances of 'anakoluth' : 'contamination' and 'Drehsatz' ('switch sentence').

5.2.1 Contamination

(57) und (58) give examples of 'contamination':

(57) gelegentlich da sind wir ins wirtshaus sind wir ja auch alle heiligen zeiten einmal gegangen (see [Wei:75:70])

(58)   ... das hilft ja wieder den Menschen, sich auch ganz darüber
        klarzumachen, was es heißt ... (see [Rat:75])

'Contaminations' like those can be reduced to a common technique, i.e. reducement. In a contamination *abcd* the chain *abc* represents a wellformed onset, and *abd* represents a wellformed sentence.

5.2.2 Switch Sentence

Also the 'switch sentence' mirrors the above mentioned structure.

(59)   ob da überhaupt die Ehe als das Ursprünglichste und Natürlichste der
        Welt die Ehe geführt wird

It differs from the 'contamination' in that *b* is repeated in *d*. The exact structure therefore is *abcbd* where abc represents the false start and *acbd* the normalised chain.

5.3 Agreement mismatches

Among agreement mismatches we count bugs in the morphological markings of gender, case, number and person as well as the choice of a wrong preposition. Those bugs, however, proved to be not very common in the DICOS and FACID data. Not regarding the dialectal variations that can be described regularly, German speakers mostly get their morphology right. For speech understanding, however, this area is virulent. The speech recognition modules of today are not reliable as to what the grammatical endings of the words are and it is up to the understanding components to match their faulty output with the strict regularities of their grammars.

6   Conclusion

This article listed the most important properties that are specific for the syntax of spoken language.

Quantitative aspects show a discrepancy to written texts in the preferred use of short sentences, small amount of subclauses and a small use of prepositional attributes, participal constructions and infinitival constructions. In oppostion to that adjectival attributes are abundant.

Besides those quantitative aspects several constructions in the periphery of the grammar were outlined that are specific for spoken language. Among those count: 1) various sorts of introducing particles and tags, 2) 'prominence structures' to the left and right, 3) ellipses in the topic position 4) taglike parentheses 5) several sorts of discontinuities. Detailed linguistic descriptions of these phenomena are available. This gives hope that a comprehensive enlargement of existing grammars in text understanding may produce satisfying results in the analysis of these constructions.

Finally spontaneous speech was shown to be strongly characterized by corrections of ungrammatical onsets by the speaker himself.

7 Bibliography

[1] [Alt:81], Hans Altmann, Formen der "Herausstellung" im Deutschen. Niemeyer, Tübingen,1981.

[2] [Aue:91a], Peter Auer, Vom Ende deutscher Sätze. ZGL 14,1991.

[3] [Aue:91b], Peter Auer, Zur Verbspitzenstellung im gesprochenen Deutsch. Vortrag DGfS-Jahrestagung 1991, Aachen, unpublished ms.,1991.

[4] [Bay:7], Klaus Bayer, Verteilung und Funktion der sogenannten Parenthesen in Texten gesprochener Sprache. Deutsche Sprache 1:64-115,1973.

[5] [Beh:27], Otto Behagel, Geschriebenes und gesprochenes Deutsch. Von Deutscher Sprache, 1927, p. 11-34.

[6] [DIC:1], DICOS, Protokolle und Transkripte DICOS-Bahnauskunftssystem. Teil 1, System 1, gesprochener Input, Universität Regensburg.

[7] [DIC:2], DICOS, Protokolle und Transkripte DICOS-Bahnauskunftssystem. Teil 1, System 2, gesprochener Input, Universität Regensburg.

[8] [DIC:3], DICOS, Protokolle und Transkripte DICOS-Bahnauskunftssystem. Teil 1, System 3, gesprochener Input, Universität Regensburg.

[9] [DIC:4], DICOS, Protokolle und Transkripte DICOS-Bahnauskunftssystem. Teil 1, System 4, gesprochener Input, Universität Regensburg.

[10] [Eng:74], Ulrich Engel, Syntaktische Besonderheiten der deutschen Alltagssprache. in: gesprochene Sprache, p. 199-228, Schwann, Düsseldorf,1974.

[11] [Gro:68], Siegfried Grosse, Mitteilungen ohne Verb. in: W. Besch (ed) Festgabe für Friedrich Maurer, Düsseldorf,1968.

[12] [HFM:81], Karl E. Heidolph and Walter Flämig and Wolfgang Motsch, Grundzüge einer deutschen Grammatik. Akademie Verlag, Berlin,1981.

[13] [Hoh:75], Christel Höhne-Leska, Statistische Untersuchungen zur Syntax gesprochener und geschriebener deutscher Gegenwartssprache. Akademie-Verlag, Berlin,1975.

[14] [Kle:85], Wolfgang Klein, Ellipse, Fokusgliederung und thematischer Stand. in: Reinhard Meyer-Hermann and Hannes Rieser (eds), Ellipsen und fragmentarische Ausdrücke, p.1-24, Niemeyer, Tübingen, 1985.

[15] [Kra:91], Jürgen Krause and Bettina Mielke and Huberta Kritzenberger, Computertalk. to appear

[16] [Rat:75], Rainer Rath, Korrektur und Anakoluth im Gesprochenen Deutsch. Linguistische Berichte 37:1-12, 1975.

[17] [Ric:66], Helmut Richter, Anleitung zur auditiv-phänomenalen Beurteilung der suprasegmentalen Eigenschaften sprachlicher Äusserungen. in: E. v. Zwirner and H. Richter (eds), Gesprochene Sprache. Probleme ihrer strukturalistischen Untersuchung, p. 11-21, 1. bis. 5. Rothenberger Kolloquium, 1966.

[18] [Ulv:74], Bjarne Ulvestad, Das pränukleare Adverbialattribut bei Nominalen im Deutschen. in: Gesprochene Sprache, p. 267-282, Schwann, Düsseldorf, 1974.

[19] [Wei:75], Andreas Weiss, Syntax spontaner Gespräche. Schwann, Düsseldorf,1975.

[20] [Wun:84], Dieter Wunderlich, Zur Syntax der Präpositionalphrasen im Deutschen, Zeitschrift für Sprachwissenschaft 3:65-99,1984.

Human Language and the Computer

Udo L. Figge
Ruhr-Universität Bochum
Universitätsstraße 150, D-44801 Bochum

1   Introduction

As an attribute of the noun *language*, the adjective *human* may appear more or
less tautological. I want to stress, however, that in this contribution I regard
language as one of the biological characteristics of a certain mammalian
species, namely of *Homo sapiens sapiens*. I do so in order to gain a point of
view that is as far away as possible from computers as a non-biological, a
technical species.

There is a threefold relation between human language and the computer:

-   The computer may assist the study of human language.

-   The computer may support or even replace human language activities.

-   The computer may be controlled by means of human language.

Although the corresponding activities usually are all termed *computational
linguistics*, there are some distinctions to be made[1]. Computer-assisted language
study clearly is a form of linguistics. But as computers allow for linguistic
research work that formerly could not be thought of (e.g. algorithmic simulation
of language production and reception processes), there may well be a branch of
linguistics engaged in the analysis and systematization of the use of computers
and the application of computer science in the study of language. - The
technical support of human language activities has had a long tradition. All
writing implements, the ear-trumpet, the overhead projector - to mention only a
few examples - belong to this tradition. The computer is just its latest and its
most powerful link. - Every technical device has to be controlled in some way

---

[1] Many computational linguists have tried to structure their field (cf. Bátori / Lenders / Putschke 1989: XV).
The difficulties in obtaining generally accepted results are not only due to the matter itself, but also to the fact
that it is almost impossible to term subdisciplines of computational linguistics (and even the discipline itself)
in a convincing manner.

or other. The computer, however, is the first machine to be controlled symbolically. This has two important consequences. First, implementing language as a means of controlling computers is real technological pioneer work, presupposing a good knowledge not only of what a computer is, but also of what language is. Second, because it is done by means of symbols, the control of computers is a matter of considerable theoretical interest above all for semiotics.

In what follows I shall begin with an outline of a biological conception of language. On the basis of this conception I then shall deal with the two forms of language technology mentioned above, viz. the automatic support or replacement of human language activities and the control of computers by means of language.

2   Human language

Human language is a particular instance of a semiotic system. *Semiotic systems* of different kinds are widely spread over the animal kingdom. They consist in a faculty of organisms enabling them to perform two countercurrent processes:

1. the manifestation of inner states that are not perceptible to the environment of these organisms by the production of signals that appear on their surface and, therefore, can be perceived,

2. the modification of such inner states in consequence of a specifically semiotic[2] processing of the perception of such signals.

---

[2]   Instead of *specifically semiotic* I often use the term *oblique.*

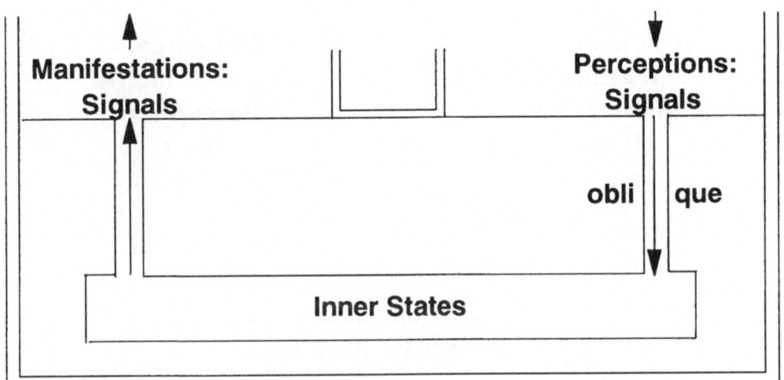

Figure 1: Semiotic System

The inner states can be of three different types: motivational (e.g. mating motivation), emotive, and cognitive. The manifestations are mostly movements. The signals produced can be perceived by one of four types of sensory modalities: a chemical, a tactile, an optical, or an acoustic one.

Language is a special instance of a semiotic system in different respects: The inner states manifestable or modifiable by language are clearly cognitive. They are chunks of knowledge or, as I prefer to say, sections of memory[3]. That is to say that they have the same structure as memory as a whole: They are networks of concepts and features[4]. The manifestations consist in oral and pharyngal or in

---

[3] Bátori / Lenders / Putschke put it in the following manner: "... human language is irrevocably allied with thought" (1989: XVIII).

[4] For details cf. Figge 1989.

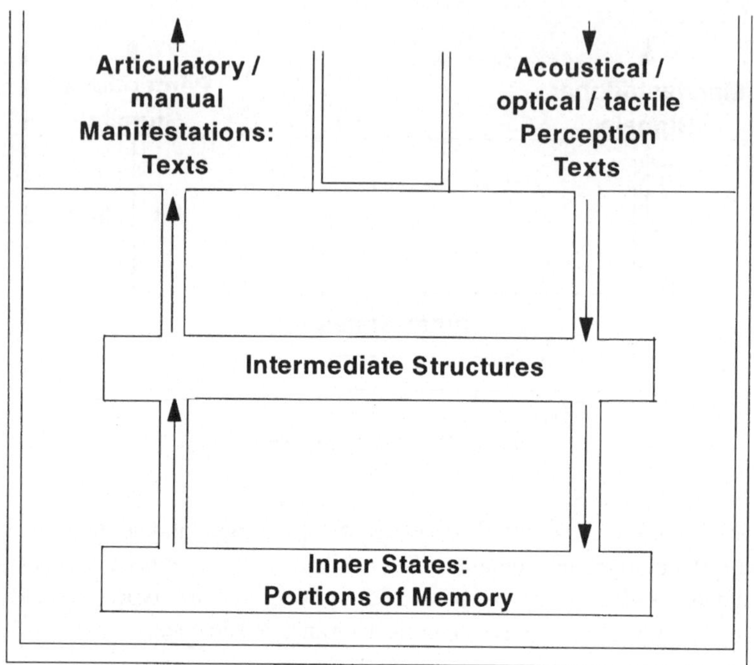

Figure 2: Human Language

manual and brachial movements. The signals are perceived acoustically (spoken
texts), optically (written or signed texts) or through the tactile modality
(Braille).

The coupling of a semiotic and a cognitive system is uncommon, but not
restricted to human language. However, language has an important feature of its
own, namely the specific syntactic structuring, the *duality of patterning*
underlying its signals. This feature is so exceptional that it requires a special
biological explanation. Most probably, syntax arose because there was no co-
evolution of memory and speech. Memory is a very old feature of animal
organisms (cf. Vogel / Angermann 1984: 387). As the human brain is
essentially a mammalian brain[5], human memory may be considered essentially,
in its basic structure, a mammalian memory. That is to say that its age is at least

---

[5]   The most important common structure is the neocortex (cf. Rauber / Kopsch 1987: 418-421).

200 million years. Man's articulatory faculties are much younger. They are only two million years old at the very most. Their development began, therefore, when memory had already matured. Furthermore, it appears that man's faculties for speech did not develop as such, as semiotic or specifically linguistic faculties, but as a mere by-product of a general reorganization of the left hemisphere of the hominid brain. Only secondarily, slowly and step by step, within an already existing multimodal semiotic system, they assumed a manifestational function[6]. As many other biological phenomena, human language is an evolutionary patchwork. One of its most striking features is the discrepancy between the linear structure of its signals and the network structure of the underlying inner states, resulting from the course of evolution. The best way to explain syntax is to consider it a sort of buffer between the network structure of memory and the linear structure of the articulatory movements. In any event, it is not - as it might seem - a gift of God to linguists allowing them to develop a new syntactic formalism every year and then to apply it to language technology among other things. It rather is a device adapting the linguistic mode of manifestation to the inner states manifested and thus depending upon the peculiarities of human articulation as well as on those of human memory. Because of the mediating function of syntax I speak of intermediate structures rather than of syntactic structures.

3 Language technology

In the introduction I distinguished two forms of language technology, viz. automatic support or replacement of human language activities and control of computers by means of language. I now shall return to this issue.

3.1 Automation of language

One could conceive of a fully automated system that duplicates a person's memory and, on this base, produces and understands texts for this person, e.g. a system that automatically deals with his or her correspondence or his or her telephone calls. Such a system would be the result of a complete automation of the language of this person. Of course, such a system is quite unrealistic at present, but it can serve as a framework for more realistic conceptions.

---

[6]   For details cf. Figge 1990.

What seems realistic, is a text processing system that integrates all existing aids for *text production* in order to support all the steps of this process. In particular, it should include

- a component supporting the structuring of thoughts with a view to facilitating the exposition of a given topic,

- a syntactic component functioning as a checker or as a corrector of the text supplied by the user, which could, however, be developed into a sort of adviser offering syntactic patterns for the manifestion of previously chosen conceptual configurations,

- a lexical component checking or correcting the words supplied by the user (including their spellings), which could also assume advisory functions, offering words for the ideas to be expressed.

Such a system would be an effective automatic aid for text production. In particular, it would be a very helpful tool for the production of foreign-language texts.

As to the automation of *text understanding*, it is not so easily conceivable how it could be done and what it would be good for. Of course, there are systems that have to do with text understanding - systems for automatic indexing, abstracting, and content analysis -, but they do not support reading as directly and personally as text processing systems support writing: They rather prepare texts with a view to saving reading time. To this extent, of course, they contribute to support linguistic activities.

There are linguistic activities that are less common, more professional than text production and understanding, namely *translating* and *language teaching*. The attempts at automation in either field are well known. It should, however, be mentioned that not only language teaching, but teaching in general is heavily dependent upon language. Thus not only computer-assisted language teaching (CALL), but computer-assisted instruction (CAI) in general is a matter concerning language technology.

An aspect of language that has not yet been taken into consideration in this section is the plurimodality of its manifestations: acoustic, optical and tactile. There are six possibilities of *transforming a signal* of one modality into a signal of another one. Three of them are of practical interest, mostly for handicapped people:

1. acoustic → optical: transformation of spoken into written or into signed texts,

2. optical → acoustic: transformation of written or signed into spoken texts,

3. optical → tactile: transformation of written into Braille texts.

A device for transforming spoken into written texts is the automatic typewriter or talkwriter, which is still under development. It has been conceived for the office, but it could also be used by the blind. There are also different kinds of systems for the automatic transformation of written into spoken texts. They are used for announcements and the like and as reading machines for the blind. The development of a system that would automatically transform signed into spoken texts and vice versa would be a real challenge (at least as fascinating as the automatic analysis of traffic scenes and their automatic verbal description). But as only one per mil of a population is deaf-mute and not all of them know a sign language, probably the market would be too small.

3.2   Control of computers

The computer differs from other machines in having been designed for information processing, as a machine supporting or taking over mental rather than physical work. Therefore, it appears quite natural that from the very beginning computers have been equipped with symbolic rather than physical control devices. Symbols or other kinds of signs or - to put it in the terms introduced above - *semiotic systems* are indeed the most adequate means of dealing with information.

In the beginning, the semiotic systems implemented in computers exclusively had the form of *command languages*. Lexically and syntactically, command languages correspond in a fairly direct manner to the inner states underlying them, viz. objects, relations and functions. Their similarity to human language is restricted to the fact that their signals include letters and that their words may show some resemblance to human words for mnemotechnic reasons. In the main, however, they must be learned as a special kind of formalisms.

As this is a laborious task, other semiotic systems have been designed for computers, in order to make their control more easy, namely *menu* and icon or other *graphic* systems. Such systems are much more straightforward than command languages, but share with them their laconism.

Furthermore, there are attempts to make *human language* a means of computer control. These attempts pose two kinds of problems, one concerning the pragmatics, the other the structure of human language. The pragmatic problems result from the fact that, as human language is coupled with memory, its main function is the exchange or, to put it more realistically, the intruding and the gaining of knowledge. Even social control frequently has the form of knowledge manifestation. Thus, penal laws do not forbid anything: they just describe the consequences of certain acts. There are, of course, linguistic means that support social control more directly, e.g. imperatives. There are, above all, sublanguages that function as human command languages, e.g. military or nautical command languages. These are languages for special purposes, characterized by a meager syntax and a largely standardized vocabulary. In this respect they are similar to computer command languages, but they differ from them in being far less artificial. So, one could say that implementing human language in computers could simply consist in making the signals of command languages less artificial, more similar to those of human language[7]. Thus, they would be accepted by the users as readily as human command languages are accepted for instance by corporals or captains.

This would be true, if command languages were unrivaled as control systems. As a matter of fact, nowadays they have to compete with menu and graphic systems. Now these systems have an advantage that command languages scarcely ever will have. While command languages are true control languages in that they leave the initiative to the user, menu and icon systems take the initiative offering the user a limited number of alternatives. This evokes a feeling of security that even less artificial command languages never can produce. So it is no wonder that natural language plays only a minor part in computer control.

For whatever purpose a computer may be used, it is a machine that has to be controlled. The users, however, may assume other attitudes towards it. This is true above all for systems that include some knowledge base and allow for a kind of knowledge exchange with the user, e.g. question-answering systems. In this case, there is a semiotic parallel between the system and the user. Both manifest, on their surface, knowledge by means of signs. So the user may consider the computer a sort of dialogue partner rather than a tool to be

---

[7] DOS-MAN seems to be such a deartificialized command language.

controlled. In consequence, he might find it quite natural to talk with the computer in his own language. Thus, the interaction between men and computers including some knowledge base seems to be an important field for an extended implementation of natural language in the computer (cf. Bunt 1988).

On the whole, whether human language can be adapted to the control of computers and the extent to which this can or should be done - these are pragmatic questions, having to do with expectations and attitudes towards the machine.

As mentioned above, for systems such as question-answering systems natural language could be a convenient means for the communication with the users. But what does this mean? Most computer linguists think that a natural language system has to reproduce human language in detail, that it has to have different components, especially a syntactic and a semantic one, with rules corresponding closely to those of human language, and that its main task is to transform the signals put in by the user into semantic representations, which then are mapped onto the underlying formalism, e.g. a data base, by means of a special query language. Such a system can be characterized as a kind of homuncule which consults the knowledge base instead of the user. Another way of designing natural language systems would be to take into account only the essential features of human language, especially the role of syntax as a buffer between two originally independent structures, viz. the essentially linear structure of the signals and the network structure of memory. This would mean that only the signals put in by the user and put out by the computer would be language-like, whereas the syntax would have to be created independently of human syntax, with a view to making the way between the signals and the underlying knowledge base as short and uncomplicated as possible. To give a fragmentary example: A relational data base is a sort of matrix. A semiotic system with language-like signals that is coupled with a relational data base should include a syntax working with two main word classes, viz. words for rows and words for columns and establishing syntactic relations between the members of either class. All other word classes would have to be defined with regard to these main classes and will probably include a class of neglectable words[8].

---

[8] A simpler example is the syntactic analysis of semi-artificial natural-language expressions as a first step towards their translation into expressions of a command language. In this case the translation system could be considered a kind of semiotic system with the natural-language expressions as signals and the commands as inner states. A typical UNIX command, for instance, consists of a command name, an option and one or more arguments (cf. Gulbins 1988: 34-35.). So the syntactic analysis of natural-language expressions has to be

4   Linguistics and language technology

As has been pointed out, language technology has two different aims: automation of language on the one hand, implementation of language as a means of computer control on the other hand.[9] Automation of language can hardly be done without the help of linguistics, above all in its computerized form. In order to design systems for the automatic support or replacement of language activities, large machine-readable lexical, morphological and syntactic descriptions of a lot of languages are needed, which can only be supplied by linguists, especially computational linguists. For the design of linguistic control formalisms for computers, however, a recourse to results of linguistics as the science of human language could be helpful in general, but futile or even harmful in detail. Computers represent a separate species. They have to be supplied with species-specific semiotic systems, even if these systems are means of inter-species communication with humans.

Natural language systems for man-machine communication can adequately be dealt with only in connection with other semiotic systems designed for this purpose. All such systems form the field of research of an emerging subdiscipline of semiotics, viz. computational semiotics[10].

5   References

[1] Andersen, P. B. (1990): A Theory of Computer Semiotics. Semiotic Approaches to Construction and Assessment of Computer Systems (Cambridge Series on Human-Computer Interaction 3). Cambridge: Cambridge University Press.

[2] Bátori, István S. / Lenders, Winfried / Putschke, Wolfgang (eds.) (1989): Computational Linguistics. An International Handbook on Computer Oriented Language Research and Applications (Handbücher zur Sprach- und Kommunikationswissenschaft 4). Berlin / New York: de Gruyter.

---

done with a view to identifying words for command names (a kind of verbs), words for options (a kind of adverbs), words for arguments (a kind of nouns) and irrelevant words. In the expression
Show me all files!
show is a word for the command name 'ls', *all* a word for the option '-a' (and by no means a quantifier or a determiner), *files* a word for an (empty) argument and *me* an irrelevant word (as irrelevant as the exclamation mark). Syntactically, the word *all* has to be analyzed as governed by *show* rather than by *files*.

[9] For a more extensive discussion of this topic cf. Heyer / Figge 1989.

[10] Cf. Andersen 1990, Figge 1991.

[3] Bunt, Harry C. (1988): 'Natural Language Communication with Computers: Some Problems, Perspectives, and New Directions'. In: van der Veer, Gerrit C. / Mulder, Gijsbertus (eds.): Human-Computer Interaction: Psychonomic Aspects. Berlin etc.: Springer, 406-442.

[4] Figge, Udo L. (1989): 'Gedächtnis, Wissen, Denken, Sprache'. In: Becker, Barbara (ed.): Zur Terminologie in der Kognitionsforschung. Workshop in der GMD, 16. - 18. November 1988 (Arbeitspapiere der GMD 385). Sankt Augustin: Gesellschaft für Mathematik und Datenverarbeitung mbH, 27-38.

[5] Figge, Udo L. (1990): 'On Language Origins. Consonants and Cognition', In: Koch, Walter A. (ed.): Geneses of Language. Genesen der Sprache. Acta Colloquii (Bochum Publications in Evolutionary Cultural Semiotics 11). Bochum: Brockmeyer, 200-227.

[6] Figge, Udo L. (1991): 'Computersemiotik'. In: Zeitschrift für Semiotik 13, 321-330.

[7] Gulbins, Jürgen (1988): UNIX. Eine Einführung in Begriffe und Kommandos von UNIX-Version 7, bis System V.3. Dritte, überarbeitete und erweiterte Auflage. Berlin / Heidelberg / New York: Springer.

[8] Heyer, Gerhard / Figge, Udo L. (1989): 'Sprachtechnologie und Praxis der maschinellen Sprachverarbeitung'. In: Sprache und Datenverarbeitung 13/2, 41-51.

[9] Rauber / Kopsch (1987): Anatomie des Menschen. Band III. Nervensystem und Sinnesorgane. Ed. by Leonhardt, H. / Töndury, G. / Zilles, K. Stuttgart / New York: Thieme.

[10] Vogel, Günter / Angermann, Hartmut (1984): dtv-Atlas zur Biologie. Tafeln und Texte (dtv 3221-3223). München: Deutscher Taschenbuch Verlag.

System Architectures for Speech Understanding and Language Processing

Walther v. Hahn and Claudius Pyka
University of Hamburg
Computer Science Department
- Natural Language Systems Division -
Bodenstedtstraße 16, D-22765 Hamburg
{vhahn,claude}@nats2.informatik.uni-hamburg.de

1   Introduction

Processing natural language dialogues obviously is much more than simply processing a sequence of more than one sentence. Similarly, processing spoken dialogues is much more than handling just another type of input.

Our working hypotheses (relying at least on minimal cognitive claims) are, that integration of speech and language must be incremental, synchronous and more or less deterministic. In other words, processing must

- incrementally follow ongoing signal interpretation of low level recognition modules (in contrast to sentence oriented processing), even though the relevant units to be analysed on each level are of different size,

- synchronously proceed with the ongoing reception of the speech signal such that processing time and elapsed real time always are in a defined linear relation to each other (ideally 1:1)

- deterministically end up with one unambiguous interpretation of the speech input without resort to backtracking or the need to reanalyse previous intervals.

These principles first of all require a much more elaborated architecture for the interconnection of all the modules on all levels and not only another (spoken language) lexicon and some more specific modules. A sequential architecture where modules follow each other in a fixed sequence (say in accordance to a phonetic/linguistic layer model) and work as filters on the set of hypotheses from the lower level, will not meet the principles explained above.

This paper describes the principles and decisions concerning the system architecture within the project ASL[1] (Architectures for Integrative Speech and Language Processing). We do not report on all those activities in the project which contrast parallel and connectionist architectures[2] to explicit software architectures.

Two examples will sketch the requirements of such a speech/language system:

2   Two examples

A system must come along with the following two types of input which are admittedly the right and left extremes on the scale of interaction.

(1) (what was his name?) "Zenon Pylyshin"

Recognition of names cannot rely on any semantic or pragmatic knowledge, the information can only be extracted from the signal itself. There are (at least for the last name) no expectations or regularities. In this case, the linguistic modules will not contribute to the recognition task (except the wider context). Opposite to this "no-linguistics, signal-only"- case recognition, the next example relies on semantics only :

(2) Imagine the following situation:

Person A is sitting behind her desk when B opens the door and asks A: *"mbmmmbmb"* (something inaudible). A answers "Two with with". B returns after 5 minutes with two cups of coffee with creme and sugar. As everybody knows he is in charge of coffee today[3].

Signal processing will not deliver any useful result and in fact it was not necessary for person A to know whether it was "How about coffee?" or "As yesterday?" or even "How many cups; with creme or with sugar?"

The first case is easily to be handled with traditional serial methods, because these methods will always start with the signal. In the second example any signal processing is superfluous, because only the variables <number> and <ingredients> are to be instantiated. Everything else is bound by the situation.

---

[1]   funded by the German Ministry of Research and Technology under contract Nr. 01 IV 101A

[2]   Wermter92

[3]   Example from Tom Wachtel (Canon Research, London) by personal communication.

An architecture must be flexible enough to succeed with both examples. It cannot be defined beforehand for processing top-down or bottom-up or any mixture of it.

This holds especially when natural dialogues are in question. Consider the following example from a train schedule scenario:

```
A1:      Reise(?..............?) Grüß Gott!
B1:      Grüß Gott, ich möchte morgen nach Ulm fahren so gegen
         14 Uhr, könnten Sie mir da
A2:      gegen bitte?
B2:      gegen 14 Uhr möchte ich nach Ulm fahren
A3:      gegen 14 Uhr ab Nürnberg um 13 Uhr 58
B3:      Ja
A4:      dann sind sie in Donauwörth 14 Uhr 53
B4:      Ja
A5:      gehts weiter 15 Uhr 9
B5:      15 Uhr 9 ab Donauwörth ja
A6:      Ja in Ulm 15 Uhr 57
B6:      15 57 und einen Zug später?
```
and so on[4].

The example is a jungle of seemingly unconnected and ungrammatical utterances and meta-utterances (esp. A2, which is something like "*about, pardon?"), but everybody can understand the exchange of information.

3   Relevance of the concept of architecture

First there is a software-issue:

Obviously there exist specific architectures for specific tasks and one can judge which architectures are more suitable for a given task and which are less. The thorough choice of a suitable architecture for a given problem makes the solution of a problem easier (in terms of time and ressources). And: Architecture is more than the local flow of control and more than interface definitions.

Secondly, there is a linguistic issue:

---

[4] Material provided by J. Krause, Regensburg University.

The order of the processing modules and their processing steps is a partial model of the human communication processes. A language theory has to address the linguistic aspects, the speech signal aspects and the process aspects (including time issues).

> *"It is equally necessary to specify in what sense linguistics is about processes in human beings and to clarify the nature of these processes and their relation to the systems of sentences, their structures and meanings, i.e. to language in the abstract sense."* [5]

4   A closer look to processes

Let the following state of recognition in a combined speech&language processing system be reached:

NL input: System: Sie möchten nach Nürnberg reisen?
             (You want to travel Nürnberg?)

Client: ... nach Nürnberg ... Ja!
             (to Nürnberg .... Right)

---

(1)   The Parser is processing the client´s reply, with the following existing word hypotheses to be checked:

{nach Bamberg, Hachenberg, nach Nürnberg, Bach den Berg, macht nur er}

additionally the parser fails to find an arc for the following input: *Ja.*

---

(2)   Request from the parsers to the dialogue module:

Which category is expected?

Answer: Syntax: PP (location)

          Semantics: goal (x).

---

(3)   All other hypotheses except {nach Bamberg, nach Nürnberg} can be cancelled.

---

[5]   Helmut Schnelle in ThLing 1981.

> (4) The missing information now is:
>
>   o Is this the end of the utterance? or
>   o Is an elliptical part finished?
>   o if so, is there any segmentation signal being recognized?
>
> (5) Request to Prosody: End of utterance?
>   Any segmentation signal?

> (6) Prosody sends to the parser:
>
> end of utterance = FALSE,
> segmentation signal = pause (<from>, <to>)

> (7) Result:
>
>   o "Ja" can be a confirmation of a previous phrase
>   o starting point for segment Ja! fixed
>   o "nach Nürnberg" must be the correct interpretation, because it is the only
>   item which was mentioned before and can be confirmed.

5   What does the concept of architecture include?

As a first step when considering architectures we have to decide whether the problem can be solved by a monolithic approach, e.g. by one algorithm, or whether we need a decomposition into partial solutions. The complexity of natural language understanding implies an architecture composed of several components. A "monomodular" architecture is an extreme case, normally architecture in its proper sense is just

- a theory about the existence of several functionally distinguishable modules and their cooperative interaction, directed towards the solution of the problem.

- An architecture describes

  - the internal structure of a complex system; that means
  - the components of a complex system,
  - the nature of the links between the components and
  - the way the components cooperate (what, how, when).

- The design of an architecture is one step towards a concrete solution of the complex problem.

- A software architecture consists of

    - autonomous modules with specific functionality, internal states and interfaces to the outside,
    - logical connections to other modules,
    - a language common to all or several modules and
    - a description of the flow of control between the interacting modules.

The next step after having decided about an interactive architecture is the corresponding description of all the subparts, the description of the methods to solve the subproblems, and the description of the interaction of the partial solutions to solve the whole problem. This step usually is called the development of a "Conceptional Architecture" (CA).

A third step concerns the realization of this CA in a software architecture, which is

- Mapping of the partial approaches onto the tasks of the software modules,

- choice of software techniques which control the interaction of the modules according to the CA,

- Establishing methods to recompose the partial solutions,

- Evaluation techniques.

6   Levels of abstraction

From the previous discussion it should be obvious that it is an over-simplification to speek about architecture as a phenomenon which is bound to only one level. There are at least the following five layers on which an architecture must be planned, designed, and implemented:

- 1 -	Cognitive Model Layer
- 2 -	Conceptual Layer
- 3 -	Software Realization Layer
- 4 -	Implementation Layer
- 5 -	Layer of Standards and Interfaces

Reconsidering the example in chapter 4 ("Nach Nürnberg") we can now clearly indicate, which information is available and which decisions are necessary on which layer. Let me give some examples to clarify our view of several layers:

On the first layer the pragmatic/semantic context can be described being the communicative background of the speaker, goals and strategies to achieve the goals can be defined with their processes attached. The world knowledge and the perception of the given situation are further conditions of the understanding of the exchange of both partners.

On layer - 2 - one can describe actions like the message "This is a preposition" from morphology to syntax, or "the phrase under work most probably is a spatial description" from semantics to syntax.

More details on layer - 3 - is given in chapter 9, where the realisation as communicating modules is described together with the rules of communication and the handling of messages within the components.

On layer - 4 - we decide among others to use in the UNIX environment the interprocess communication with RPC and the means of "sockets".

A prerequisite, on layer - 5 -, is the definition of a common formalism, say feature matrices, for unification among all layers. Furthermore, an interface description language must be defined[6].

All these layers are based on a sixth layer, the hardware layer, for which, however, there is no specific model in the ASL project.

On top of the layers there must be in principle another layer, where the evaluation of the models must be performed. Psycholinguistic considerations (or, maybe, other criteria such as efficiency or ergonomy) must be applied to compare different solutions and evaluate the performance.

7   Architecture and time

The interaction of modules implies considering temporal relations between these modules. This dynamic view of interacting modules in a natural language understanding system contrasts with the mostly static view of linguistic and cognitive models where the focus is put on descriptions of observable

---

[6]   Pyka1992

phenomena and their explanation using rule-based methods. Only, when dealing with phenomena like

- syntactic rule ordering which implies explicit temporal orderings of rule activation/application, or

- data analysis direction (e.g., left-right parsing) which implies a temporal order in the availability of analysis results, the parameter of time becomes an issue.

In speech processing, there always is a natural time given by the sequential nature of the incoming speech signal. In an integrated model of speech understanding this natural time of the speech signal must be taken into account when specifying interaction patterns between modules. The availability of signal features incrementally follows the incoming speech signal. An architecture for an integrated speech understanding system must specify communicative patterns among modules, explain temporal restrictions on accessability of signal data and extracted signal features, and explicate how information from separate modules may benefit the understanding process.

The lack of knowledge about temporal relations in language understanding is the main obstacle for our practical work.

8   Current models

In contrast to the architecture which is designed in the ASL project there exists a broad choice of alternatives, which are briefly summarized in this chapter.

8.1   The sequential depth-first model

The basic processing is as follows: Every meaningful segment of the signal level (according to the definition of the entities to be recognized) is processed bottom-up as soon as applicable on all higher levels (phonology, morphology, syntax etc.) in the sequence as the segments come up. The advantages of this approach are: There is always a well-defined environment of true hypotheses available on each level. Few material is held for further processing. A low software engineering effort is necessary because the structure allows for independent development of each component. There are some shortcomings of this processing, however: No revision of results is possible, all processing remains very locally, therefore no long distance influences among components can be exploited for reducing the work load of the whole system.

8.2 The sequential breadth-first model

This model lines up the processing components into a linear sequence. Each component receives its input data from a unique predecessor component and sends its results to a unique successor component. A component computes all hypotheses possible given its input data and its rule base and passes these hypotheses to its successor. Input hypotheses from its predecessor are filtered by a component. Only input hypotheses that can be locally incorporated into one of its own hypotheses are passed to the successor component, extended by additional information from the component in focus.

The advantage of this model is that each component receives all possible hypotheses from its predecessor and thus has all available information to compute the best hypotheses licensed by the input hypotheses and the rule base.

There is no direct interaction between any components. Thus it is not possible to resolve ambiguities that would require such interaction. Another disadvantage of this model is that many hypotheses are generated and passed on between components that are rejected later by succeeding components. A subtle disadvantage arises when components in principle would be able to resolve some ambiguity, but only if some additional information from a preceding component would be available. But this would conflict with the informational encapsulation of components that states that each component only deals with one kind of information, e.g., that a parser only deals with syntactic structures.

8.3 Static multilayered model

This model groups several components into a communication network that statically defines all possible dependencies between components. Each link between two components has its specific communication language defined. There is no way to dynamically change the dependency pattern during analysis. In fact, this model basically adopts the sequential organization of components and inserts additional communication connections where this seems suitable. One example is a prosodic component which extracts prosodic information from acoustic data and passes this information to the syntactic parser, the semantic analyser, and the pragmatic analyser.

The static multi-layered model avoids the most striking disadvantage of the sequential model, i.e., in this model it is possible to apply any information from any component if this allows to resolve some local data ambiguity. But the model requires complete knowledge on behalf of the model implementor who must in advance (before an implementation of the model) specify the extra links between components. The qualifier "static" stems from here.

This model is half-way towards a dynamic model where components are free to communicate with each other without restricting the actual communication pattern previous to the model's implementation. In this respect, the static multi-layered model is influencial to ASL, where general communication will be supported.

8.4   The interactive model

The interactive model is the logical extension of the static multi-layered model, allowing unrestricted communication between components. There is no preferred flow of information through the components, at least in principle. Data dependencies imply such a preferred ordering of components. In general, a component depends on data supplied by one other component. This producer component is then considered as being sequentially prior to the data consuming component. But beside these data dependencies, nothing restricts communication to these cases.

Through its unrestricted interactiveness, this model allows disambiguation of conflicting analyses by incorporation of any information available at some other component. Generation of hypotheses eventually being rejected by other components may be avoided in many cases when the respective information can be obtained interactively.

But this model in its generality lacks directionality towards an analysis of the input data. There are no principles to guide the components towards a solution, and thus it may happen that a model implementation may actually be less efficient and successful than implementations of sequential or breadth-first models. Another potential danger with this model is that communication may slow down system performance significantly if costs of analysis are not weighted against costs of communication.

8.5 The model of Briscoe (1987)

The model of Briscoe[7] is a mixture of previously discussed models, combined with some principles that are postulated for the task of speech and language processing. The departure point is a serial model where the participating components are lined up one after the other, passing their respective hypotheses sequentially. The first principle states that each component is able to analyse its data deterministically. In case of an ambiguity, a component can interactively request further information from some component that may supply such information. If additional information is received, the ambiguity is resolved respectively. Otherwise, a default selection among the conflicting analyses is done.

A second principle restricts the type of communication to simple yes-no requests. A requesting component may only ask another component whether some feature is present in the speech signal. The response is limited to a binary value, meaning yes or no, respectively, or no response at all, forcing a default selection. Responses must obey a strict temporal restriction, in that they must occur immediately or never. Requested components either have the requested information to respond to the request, or they skip the request.

It is questionable, whether Briscoe's model is extendable to a model that may account for a wide range of speech and language phenomena without being rendered too complex. Briscoe's model only centers on the morphosyntactic parser and its informational requirements. Extending the model to other components needed in a speech-language system may show that there must be too many interactive communication channels which would result in a general interactive model. Furthermore, it may be necessary to extend the bandwidth of these channels to transfer more than just binary information items.

8.6 The blackboard architecture of HEARSAY-II

The blackboard architecture of HEARSAY-II is a very general model[8] which was initially applied to speech understanding. A blackboard system consists of a global data structure, the blackboard, and several processing components that are called knowledge sources. Knowledge sources process data they retrieve from the blackboard in an autonomous manner, i.e., there is no direct interaction between knowledge sources. Two global processes supervise and control the

---

[7] Briscoe 1987

[8] Erman et al.

activities of knowledge sources. The blackboard monitor peeks at all blackboard accesses of knowledge sources and registers blackboard events that might trigger knowledge sources waiting for such events. The scheduler controls activation of knowledge sources and assigns computing resources to knowledge sources.

The blackboard architecture was intensively investigated in the past 15 years and applied to a huge variety of different problems. The blackboard model is very flexible and general and may lead to inefficient systems, if not restricted by additional principles that guide the behaviour of the system. Further on, the central position of the blackboard introduces a real bottleneck and renders parallelism among knowledge sources difficult. The same problem for potential parallelism is caused by central scheduling.

In ASL we consider the blackboard architecture to be too general and unrestricted to be successfully applicable for integrated speech-language processing. The central positions of the blackboard and the scheduling process are bottlenecks which do not reflect the additional knowledge which is available for speech-language processing. Direct interactions between modules are more efficient, as well as distributed control.

8.7.   The architecture of DIAL (CRIN/INRIA) 1989

DIAL[9] is characterized as a natural language task oriented spoken dialogue system. Its aim is to engage in a dialogue with a user to achieve some task. The system architecture of DIAL strongly emphasizes the cooperation of specialized system components. The cooperating components in DIAL are the syntactic and semantic analyzer, the dialogue component, and lexical, acoustic-phonetic, and prosodic components.

There are two basic types of communication events between components, validation requests and local data transmissions. A component sends its locally computed hypotheses to another component for validation. The validating component responds with a rejection of a hypothesis, or with a corroboration of a hypothesis. This type of communication can be classified as strongly interactive and is similar to the request-response patterns in Briscoe's model. Local data transmission merely means passing local analysis results to interested, i.e. potentially consuming components.

---

[9]   Mousel et al 1989

The syntactic-semantic component and the dialogue component may operate in a bottom-up as well as in a top-down manner. The bottom-up mode of analysis consumes input data from previous analysis stages and computes local hypotheses of these received data. The top-down mode of operation enables a component to deduce promising interpretations of yet not seen input data, given suitable contextual clues.

The strength of DIAL is its inclusion of interactive communication events which are used for validation of locally computed hypotheses. This enables a component to early detect local hypotheses that cannot be sustained by other components. Further processing on such rejected hypotheses may be then avoided, thus saving resources for more promising hypotheses.

But DIAL does not step far enough into the potential advantages of interactive communication. It restricts validation requests to only very few cases. Furthermore, validation requests are a too restricted class of interactive communication events, excluding more elaborate forms of interactive communication. Such more elaborate interactive communication patterns are one of the topics of the ASL project.

8.8. Marslen-Wilson´s Cohort Model

The basic assumptions of Marslen-Wilson´s Model are based on the "early recognition" discovery: Three processes, namely access, selection, and integration, are defined.

> "Within the first 100-150 ms the recognition devices corresponding to all of the words in the listener´s mental lexicon that begin with the initial sequence will become active (word-initial cohort). This subset of active elements monitors the continuing sensory input and the compatibility of the words that the elements represent with the available structural and interpretative context."

A mismatch will cause elements to drop out.

The relevance for ASL mainly consists in the concepts of multiple access, multiple assessment, and real-time efficiency:

> "These requirements, ... , cannot be met by a serial process moving through the decision space one item a time. They point, instead, to some form of parallel or distributed recognition model. But they do not, however, uniquely determine the form of such a model."

The technical difficulty with these ideas is the fact, that there is no paradigmatic implementation of the results, which exhibits the concrete software decisions and which can be compared directly with other architectures.

9   Architecture of ASL

The architecture of the ASL[10] system is based on the interactive incremental model of computation[11]. The analysis of a speech signal is performed by a couple of specialized components. The components are connected by a communication subsystem that permits the exchange of messages between any components. Message passing is the only modality of interaction between components.

Each component has an identical top-level structure which is as follows:

The task specific kernel of a component is its Processing Unit (PU). The PU applies its analysis algorithms to external data received within messages. For example, the PU of the syntax component consists of chart parsing operations and the chart data structure. The PU interacts with other components in two ways:

- directly through the Communication Interface, and

- indirectly through the Hypo Space.

The parameters in the Parameter Block influence the behaviour of the PU.

The Communication Interface (CI) is the only gate to the environment of a component. All communication is via the CI. Communication is realized by exchanging messages between components. Messages are strictly defined formal structures representing data to be passed over, and contain an explicit representation of the intended purpose of the transmission. Thus, messages are described by a syntactic structure, bear semantic contents, and express pragmatic intents.

The CI is an autonomous part of a component which works concurrently to the Processing Unit. Only if the PU directly communicates with another com-

---

[10] The project ASL has two sub-projects. The "northern" part investigates innovative interactive architectures, wheras the "southern"-part tries to optimize standard techniques. The ideas expressed in this paper belong to "ASL-North".

[11] described in more details in Pyka 1991

ponent, the CI temporarily becomes a slave to the PU passing a message to or from the environment. But the CI has additional tasks to perform. It must manage the Hypo Space and update the parameters in the Parameter Block.

The Hypo Space (HS) is a repository for all hypotheses of a component. There are two types of hypotheses, external and internal hypotheses. The relative source of the hypothesis is what makes the difference. A hypothesis generated outside the component by another component is called an external hypothesis and enters the HS through the CI. A hypothesis is transmitted between components by means of a message. The pragmatic intent of such a message is to supply a hypothesis from a sender to a receiver and is encoded by the symbol SUPPLY_HYPO.

An internal hypothesis is generated by the PU itself and placed into the HS. The CI selects hypotheses from the HS and transmits them to other components. This represents the indirect mode of communication: PUs do not transmit hypotheses directly to other components, but mediated by the HS and the CI. This mediated hypothesis transmission requires autonomous processing of the CI. The CI must select accumulated internal hypotheses from the HS in time and transmit them to other components as needed. Selection of hypotheses by the CI is influenced by the parameters in the Parameter Block.

The Parameter Block actually consists of four parameters:

To explain the meaning of the parameters, we must first take a closer look at the structure of the hypotheses. The kernel of a hypothesis is the analysis of a PU of some input data. Depending on the processing level, this kernel may contain a syllable hypothesized from phonetic segments, a word hypothesized from syllables, or a syntactic structure hypothesized from words and other syntactic (sub-) structures. Each hypothesis is annotated with a time stamp consisting of the start and end time of the hypothesis kernel. That is, the time stamp specifies that interval within the speech signal over which the hypothesis kernel makes a proposition. A second annotation of a hypothesis kernel expresses a confidence value for the hypothesis.

Now we can return to the parameters of the Parameter Block. The CI selects internal hypotheses from the HS and transmits them to other components. The CI may use different selection strategies. All strategies are influenced by the parameter values. The Sending Threshold is a kind of hard limit for the CI. Any hypothesis in the HS having a confidence value above the Sending Threshold must be selected by the CI and immediately transmitted. The rationale behind this parameter is that if its value is reached by the confidence value of some hypothesis, the respective hypothesis is very plausible and should be transmitted to its "consumer" components in any case.

The other parameters are temporal and depend on each other. The Local Signal Time is the end time of that external hypothesis that furthest extends in time, i.e., the end time of that hypothesis which has the largest value among all external hypotheses. Thus, the HS contains external data up to the Local Signal Time. The Local Signal Time is continually updated whenever new external hypotheses arrive that have external end times greater than the Local Signal Time. Thus the Local Signal Time gradually increases, but cannot decrease.

A PU consumes the external hypotheses by incorporating them into its internal hypotheses. In general, there are many external hypotheses in the HS which are awaiting processing by the PU. In what follows, we refer to that external hypothesis that is already incorporated into an internal hypothesis and whose end time is maximal among all consumed external hypotheses. The difference between this time point and the Local Signal Time is called Actual Temporal Delay and is regarded as a measure for the lag of the PU behind the most actual input data. Each PU is supposed to adjust its operations such that the Actual Temporal Delay is minimized. Evidently, this cannot be the only influence on the operations of a PU. Of equal importance are the prospects of analyses that determine the operations of a PU.

The Maximal Temporal Delay is again a hard limit for a HS. In some sense, it specifies the "memory capacity" of a component. If the difference between the end time of a hypothesis and the Local Signal Time becomes greater than the Maximal Temporal Delay, this hypothesis is "forgotten" and removed from the HS. Thus, such a hypothesis becomes inaccessible for the PU and the CI. Only hypotheses with end times within the interval delimited by the Maximal Temporal Delay are visible and may be used for analysis by a PU.

The Temporal Delay parameters have been introduced to support one envisaged feature of the architecture, time synchronous co-operation of components. Each component strives to consume its temporarily most actual input data such that the whole system continually proceeds forward in analysing the incoming speech signal. Ideally, the system would have a partial analysis of the speech signal at each time instance. Immediately after accepting the last part of the speech signal, the system would respond with an interpretation of the signal.

Another envisaged feature of the architecture also benefits from the Temporal Delay parameters and their intended use, the interactive modality of co-operation among components. Up to now we have only seen SUPPLY_HYPO type messages, i.e. messages that transport hypotheses from a sender to a receiver. There is an additional message type pair that is used for direct interaction between components, more specific, between PUs of components. One consequence of the local view on processing of a component is that the component cannot resolve ambiguities that arise locally. To resolve such ambiguities, non-local information is necessary. This non-local information may exist in other components. To benefit from this information and to be able to do ambiguity resolution, a component may "ask" other components to supply this information. The result of such requests would be dramatic reductions in search space. A component may obtain locally not available information from another component by sending a request to that component specifying the needed information. In case the requested component has the requested information at its disposal, it responds to the request by a SUPPLY_DATA type message which embodies the information.

The requests we have in mind are such that a component wants to know whether some feature of the speech signal holds for a specified speech signal interval. A component which in principle is able to answer a request because it is part of its task to compute the requested feature, must have reached the respective interval in the speech signal before it may answer the request. If time synchronous co-operation of components is enforced, it is highly probable that a component can immediately respond to a request, because it just had analyzed the speech signal interval and computed the requested feature.

10　Perspectives

10.1　Implications of the paradigm of incremental analysis

General idea: Modules transmit partial results to other modules during processing. Speech signals contain very few clues about phrase boundaries that can be recognized with high confidence. A module may delay transmitting its hypotheses to other modules and thus obstruct their processing, if only completed analyses are transmitted.

Advantages: Incremental analysis promotes parallel processing among modules. Module A enables module B to start processing by passing partial results to it, contrasted with the processing delay if only completed analysis results are transmitted.

Consequences:

- Increasing the number of modules increases integration

- Communication between modules is intensive

- Communication channels must be flexible

- Communication channels must allow high data rates

- Feedback communication must be provided

- The system directly benefits from hardware parallelism

10.2   Implications of the paradigm of deterministic analysis

General idea: Each module only releases analysis results that are unique for each signal interval. There are no competing hypotheses given away by a module. A module cannot backtrack over analysis results, once these results were transmitted to other modules.

Advantages:

- Synchronization between modules is easier

- There is cognitive plausibility for deterministic analysis

- It is economic and effective

Consequences:

- Postponing decisions must be possible

- Any available non-local information must be made accessible to aid in determining deterministically analysis results

- A limited capacity memory can be used by a module to assess competing actual analysis results

10.3   Implications of the paradigm of synchronous analysis

General idea: All modules are synchronized by the input data of the system, i.e., the speech signal. Synchronously to the incoming signal, all components proceed with their processing. Consequently, all components always analyse the identical signal interval.

Advantages: Interactive communication between modules for ambiguity-resolution is facilitated, because requests for signal features can be answered with high likelihood. This follows from the fact that the interacting components analyse identical portions of the speech signal.

Consequences:

- All modules always focus on identical speech intervals (in principle)

- Requests for signal features can be responded to immediately

- Most recent analysis results are passed with high priority

- An internal system clock synchronizes all modules. The ultimate goal is to identify this internal clock with real time.

11 References:

[1] Briscoe, E.J.: Modelling Human Speech Comprehension: A Computational Approach. (Ellis Horwood, Chichester, 1987)

[2] Erman, L.D., Hayes-Roth, F., Lesser,V.R., Reddy, D.R.: The Hearsay-II Speech-Understanding System: Integrating Knowledge to Resolve Uncertainty. (ACM Computing Surveys, Vol. 12, No. 2, pp.213-253, 1980

[3] Mousel, P., Pierrel, J.M., Roussanaly, A.: Cooperation and Representation of Syntactic-semantic and Pragmatic Knowledge in a Natural Language Task Oriented Spoken Dialogue System. In: Proc. European Conference on Speech Communication and Technology, Paris, Sept. 1989, pp. 183-186.

[4] Pyka, C.: Deterministische, inkrementelle und zeitsynchrone Verarbeitung und die Architektur von ASL-Nord. Report ASL-TR-22-91/UHH. Hamburg 1991

[5] Pyka, C.: Schnittstellendefinition mit ASL-DDL. Report ASL-TR-42-92/UHH. Hamburg 1992.

[6] Schnelle, H.: Editorial to "Theoretical Linguistics" Vol. 1, 1981

[7] Wermter, S.: Noun Phrase Analysis with Connectionist Networks. University of Hamburg, CS Department. Mitteilungen Nr. FBI-HH-205/92. 1992

Natural Language Interfaces to Data Bases: Some Practical Issues

Harald Trost
Österreichisches Forschungsinstitut
für Artificial Intelligence
Schottengasse 3, A-1010 Wien
harald@ai.univie.ac.at

Abstract

In this paper the question of what is important for the acceptability of natural language interfaces is addressed. I will argue that while linguistic coverage is an important factor for the acceptance of Natural Language Interfaces (NLIs) it is not the only one. At least as important are features which are consequences of the use of natural language as communication medium. These features can be subsumed under the terms portability, habitability and robustness. Portability concerns the ease of transferring the NLI to different hardware/software environments, different domains, different applications and different natural languages. Good habitability secures ease of use and a short training time for the end-user. Robustness is a key factor in both maintainability and ease of use of an NLI. I will discuss these features in detail and show current solutions as well as their limitations.

1   Introduction

NLIs are regarded as one of the most promising candidates to bring natural language processing power to the end-user (Bates & Bobrow 1984). But up to now the market has been rather slow in accepting the technology. Why is this so? There are lots of answers possible. Too expensive in terms of money and/or hardware/software requirements, too difficult to install and service, too clumsy to use. One obvious explanation is that they do not fulfil the high expectations of end users with regard to ease of interaction.

Is this due to the limited linguistic capabilities of the current technology in Computational Linguistics and Knowledge-Based Natural Language Processing? There exist few serious evaluation studies (e.g. Tennant 1980, Krause 1982, Jarke et al. 1985) but they seem to indicate that limited linguistic coverage is not the most important obstacle. There are a few really cumbersome problems, e.g. the treatment of discourse particles and discourse anaphora. But, by and large, the current state-of-the-art seems to be sufficient to create an acceptable linguistic behaviour in the restricted domain of database access.

What is important then to improve the acceptability of an NLI? Research in Computational Linguistics has centered around linguistic coverage. General criteria for software products are of course also valid for NLIs. But there are as well features which are peculiar for NL systems yet are not directly connected to linguistic coverage (see, e.g., Rettig & Bates 1988). Practical experience suggests that these features might be crucial for acceptance.

As an example I will use Datenbank-DIALOG (Trost et al. 1987, 1988) a German language interface to relational databases that was developed at the Austrian Research Institute for Artificial Intelligence. This system has been tested in a variety of environments and much effort was placed on user-friendliness.

2   In which contexts are NLIs the solution?

Over the years a number of interaction techniques for databases have been developed, among them graphical interfaces, menus and query languages. Natural language as interaction medium has to compete with all of these and in many contexts it will not be the best choice (see, e.g. Okden & Sorknes 1987). Let's therefore start our considerations by sketching the application areas most promising for NLIs. To do so we have to answer two questions:

- For which application areas is natural language the best solution?

- For which user profile is natural language the best solution?

To answer the first question: Suitable applications must support a large set of structurally different potential queries, otherwise standard techniques like menus are preferable. Exactly because nowadays technology is unsuited for such a situation, relatively few such applications exist at the moment. But there seems to be a large potential.

Second, there should be no need for extensive formatting of the output, because complex formatting commands are not easily integrated into a NLI. This correlates favourably with the first requirement. Answers to ad-hoc queries will probably need no high degree of formatting.

Third, queries must be easily expressable in natural language-formulas, numerical data or the like are better expressed in some formal language. One promising answer could well be the integration of NL capabilities into multi-media interfaces where the user is free to choose the most appropriate medium for his current needs.

To answer the second question: large groups of users without formal background NLIs seem to be the ideal profile, because of the short training time NLIs require. But one has to be aware that users may grow out of using an NLI over time! Another important group are infrequent users and highly fluctuating groups of users.

Lastly, there is a rather small - but influential - group of users unwilling to use formal languages, e.g. managers, doctors. The problem with them is: they have very high expectations and they will give an NLI only one chance!

3 Portability

There are several different levels of portability. In the following I will identify the most important ones and show how portability can be achieved with respect to these levels (see also, for instance, Thompson & Thompson 1985).

3.1 Hardware and operating system

Portability at this level does not pose a serious problem. If the NLI is implemented in one of the more common programming languages (e.g. the C language), portability comes at almost no cost. The problem is that these languages are not the best choice if one wants to keep development time as short as possible.

In the case of *Datenbank-DIALOG* a high degree of hardware portability was achieved by cross-compiling the original source code written in LISP into FORTRAN. This solution allowed us to have the best of both worlds: the use of a LISP machine for efficient development and the high portability of a standard language.

3.2   Data Base Management System (DBMS)

Interactive query languages like SQL are obvious candidates for a flexible interface to DBMSs. All relational databases support SQL at present or will do so in the near future. The high degree of standardization has almost eliminated the portability problem. *Datenbank-DIALOG*, for example, translates natural language utterances into SQL statements. To make *Datenbank-DIALOG* easily portable to different DBMSs - using slightly different SQL versions - the query generator forms a separate module. That way adaptation is easy.

Switching to non-relational databases changes the picture. Older types of DBMSs simply do not provide a flexible enough access to data. But exactly this flexibility seems to be necessary to make natural language access really useful. Fortunately, in the near future relational DBMSs will be the industry standard.

Concerning more advanced DBMSs like the entity-relationship model or object-oriented databases there is no comparable standard yet. But it seems to be clear that these models with their more powerful data model (see, e.g., Schur 1988) will provide new chances for efficient and acceptable NLIs.

3.3   Domain (the part of the world which is modeled)

Domain knowledge - which is specific for every application - must be kept separate from the purely linguistic knowledge represented in a universally usable grammar. Ideally, semantics should belong to the linguistic part only, while the domain model describes the pragmatics. Unfortunately, current systems rely on a drastic restriction of semantics according to the actual domain model. As a consequence, the semantics must be altered for every different domain.

In *Datenbank-DIALOG* the domain model is stored in a so-called meaning lexicon. This way there are two advantages: Syntactic regularities are represented as such, leading to a broad and consistent linguistic coverage, which remains unaltered with all applications, and the semantically oriented meaning lexicon blends well with the grammar enhancing efficiency.

Adaptation to changes in the data model are easy. The same language model may be used with different data bases as long as the application domain is the same. There must be an explicit translation step between these two models. One side effect is the fact that the natural language model will better reflect the structure of the NL queries. Since this model must be created by the user efficient data acquisition tools are necessary.

3.4 Data model (the domain model as realized in the database)

The data model describes the domain as realized in the structures of the database. The ideal solution would be to use the data dictionary of the database directly as the data model of the NLI. Unfortunately, this is not possible with the current standard in DBMSs. Typically, data dictionaries are not explicit enough to serve for this purpose and, moreover, various DBMSs come with very different data dictionaries.

Datenbank-DIALOG models the application domain in two different ways: in a linguistically motivated domain model and in a database-oriented one. The parser operates on the linguistic model when creating a semantic description of the natural language utterance. With the help of a translation table these descriptions are automatically translated to database-oriented ones. This gives more flexibility with regard to linguistic coverage because the grammar of the system need not take into account the database model which might be very different from the way facts are expressed in the German language.

The separation makes it possible to use the same linguistic domain model for different databases. Only the translation to the respective database model must be rewritten. Since the bulk of the work in adapting to a new domain lies in the creation of the linguistic domain model (vocabulary, concept hierarchy, etc.) this saves a lot of effort.

3.5 Natural language

One major drawback of the current technology in NLIs is the lack of a generic language component. With the current state-of-the-art in computational linguistics it seems impossible to design a core language system which can be parametrized to different natural languages. The consequence is that portability across languages can only be achieved by creating a different module for every natural language. To make this task as feasible as possible the - language-dependent - core linguistic component should be clearly seperable from the rest of the NLI.

4　Linguistic coverage

It is not the aim of this article to deal with problems of linguistic coverage in detail (see, e.g., Trost et al. 1987). Instead, I will concentrate on just a few points I feel to be crucial for the acceptance of NLIs.

4.1　Elliptic utterances

Contrary to the common belief that formal languages are more concise than natural language it turns out that queries posed in natural language tend to be shorter than formal queries. The most important reason for that is the possibility of natural language to omit repetitions by using ellipsis and anaphora. As a result, humans are so much used to ellipsis and anaphora that they cannot renounce their use.

Although no general solutions for the treatment of ellipsis have evolved in computational linguistics yet standard techniques to deal with important subclasses are available.

The most common technique is to take the last sentence as a pattern and to to fit in the elliptic query as one of its parts. This is easy and works quite well especially if semantic restrictions are taken into account. Take, e.g., the following example:

> *In which department does Miller work?*
>
> *Smith?*

There is a widely used class of elliptic utterances though that does not fit into that pattern. Take, e.g.:

> *In which department does Miller work?*
>
> *Since when?*

In this example the ellipsis is a constituent not present in the preceding query. The system must add it to the preceding query to arrive at a solution. Datenbank-DIALOG has a limited capacity to handle such ellipses. For all elliptic utterances which do not fall into the simple substitution class the questioned part is omitted and the ellipsis added to the rest. This heuristic covers most of the elliptic utterances occuring in database queries.

4.2 Anaphora

There are several different types of pronomina. They all share the property of standing for something refered to somewhere else in the text. Most important are personal pronouns, like:

The thief tried to escape. But the policeman stopped *him (= the thief)*.

Peter washed *himself* (= Peter).

Although there are some very difficult linguistic phenomena involved practical solutions for the limited purposes of NLIs exist. The same is true for possessive pronouns like:

Peter was looking for *his* car (= the car Peter owns).

He asked *his* wife (= the woman Peter is married to) if she could help him.

He slapped *his* head (= the head which is part of Peter) when he found out.

The situation is much more complicated with discourse anaphora like:

Peter told me about the party yesterday. *This* was a wild story.

Unfortunately, no linguistic theory is available that could be implemented in a NLP system at the moment. There are only some ad-hoc heuristics available. Clearly, this is a problem which needs to be addressed to improve the acceptability of NLIs.

4.3 Discourse particles

Discourse particles are a very important means to produce coherence and as such are widely used in all sorts of dialogues. Various ad-hoc techniques have been applied to their treatment. The most simple one is just ignoring them in the interpretation. In some cases this works but in some others it leads to awkard results.

Another possibility is to force the user not to use them. This is the way Datenbank-DIALOG deals with the problem. But quite often it makes dialogues unnatural and accordingly users feel not happy with such a restriction.

Like with discourse anaphora, there is no general linguistic account of how to interpret discourse particles semantically. There are some solutions for specific subproblems, see e.g. Saebo (1988). But much more work is needed in this area before we will arrive at an adequate solution.

4.4   Quantifiers and negation

It is outside the scope of this paper to give an account of the problems and the work in this area. Let me just make a few remarks how the domain of database querying helps in cutting the problems down to manageable size.

In Datenbank-DIALOG a version of Generalized Quantifier theory (Keenan & Stavi 1986) was implemented. This allows the treatment of all extensional natural language determiners like *seven, between 2 and 10, more than 5* including the standard logical quantifiers *for all* and *exists* as special cases. What about determiners like *some* or *many*? The solution in Datenbank-DIALOG is to allow the user to define their meaning extensionally, e.g. *some* could be defined as *more than 2, most* as *more than 50%*, and so on. Since the interpretation is user-defined it is the user's responsibility to avoid inconsistencies (the given interpretation of *most* works for finite sets only).

Negation is another one of the real hard problems in natural language processing. Fortunately, in the domain of database querying it is feasible to interpret natural language negation as logical negation. In combination with the closed-world assumption of databases a purely extensional interpretation of negation gives the right results in most of the cases.

5   Habitability

The current state-of-the-art prohibits the use of unrestricted natural language. Instead, a sublanguage must be defined. Habitability means to define that sublanguage in such a way that the user can easily determine its coverage. This can be achieved by imposing restrictions on the language which a human user finds natural.

Restrictions come in various forms on different levels. Pragmatic restrictions are determined by the contents of the database and the domain. The user must get used to the fact that an NLI has only knowledge about the application domain. No mundane knowledge can be expected.

A related problem is that the database usually models only parts of the domain - those which the designer of the database thought to be relevant. Moreover, some information is stored only implicitly. Take the feature *sex* as an example. Let's assume a personnel database where first and second names are stored but the sex of a person is not. Humans can - at least in most cases - deduce from the first name if somebody is female or male. Accordingly, if one asks "Where does Ms. Miller work?" he expects to get the answer that Mary Miller works in accounting but not that Steve Miller works for sales.

A general solution to that problem is impossible. But one should be careful to include at least such obviously missing information like the one in our example. The distinction between (linguistically motivated) domain model and data model (determined by the database) makes it easier to detect such obvious deficiencies.

Restrictions on the vocabulary seem to be inoffensive, at least if the core lexicon contains a sufficiently large basic vocabulary. More tricky are word senses. The exclusion of word senses must always be based on the pragmatics of the domain.

As stated before, limited use of ellipsis and anaphora seems to be crucial for a dialog in natural langauge. Basic capabilities to process them are therefore a prerequisite for any acceptable NLI.

Restrictions on the semantics are problematic. If a structure is syntactically acceptable for an NLI and if the semantics can be interpreted in the application domain, a query should be acceptable to the NLI. I will now give some examples where structurally similar queries need a very different semantic interpretation:

Take, e.g., comparatives:

   1) Who earns more than 10.000,-- ?

   2) Who earns more than Miller ?

   3) Who earns more than he did last year ?

The average user will not understand why 1) should be acceptable while 2) or 3) are not.

Another example are how-many-questions:

   4) How many hours does course 502 have ?

   5) How many courses does Miller give ?

In these examples the system will not only assign the same structure but also have an analogous semantic interpretation. But the pragmatic interpretation depends on the data model of the database. In 4) the number of hours will probably be stored in a column of the table *course*. In 5) the system probably needs to count the number of courses where Miller is *lecturer*. The human user will expect that both queries lead to correct answers.

Also important is the fact that a question can be posed in all the different structural forms usually accepted with the system, i.e. if passive voice is supported at all it should work with all appropriate verbs.

e.g. active vs. passive

> *Does Miller give course 502?*
>
> *Is course 502 given by Miller?*

e.g. relative clause

> *Who attends the course which Miller gives?*
>
> *Who attends the course which is given by Miller?*
>
> *Who attends the course given by Miller?*

Clearly, the main architecture of an NLI determines how easy it will be to design a habitable system. Approaches like pattern matching or simple case frame approaches lack the necessary syntactic generality. The burden to create a habitable system rests therefore on the person creating the actual application. This is not a desireable situation.

On the other hand, having a powerful syntactic parser does not necessarily solve the problem. Because, the NLI must be able to interpret syntactic structures semantically/pragmatically in a meaningful way.

Let's now have a closer look at some special problems which I feel have to be addressed in a habitable system.

5.1   Standard phrases

There are a lot of standardized phrases which are typically used by humans in dialog situations:

> *1) Could you tell me...*
>
> *2) I should like to know...*
>
> *3) Please print me...*

Such phrases do not contribute to the content of the query. On the other hand, they are often syntactically (and semantically) complex. An efficient solution is to use a syntagmatic approach (pattern matching) to process them. We then need an interface to transmit the relevant information to the main analysis because syntactic (and semantic) information must be transmitted.

In 1) and 2) the rest of the sentence is a subordinate clause, in 3) a noun phrase. In this last example we also must regard the request for printing a hardcopy instead of using the default output medium.

5.2 Handling of proper names

An NLI should not require that proper names are explicitly entered into the lexicon because:

- a database may contain a very large number of names,

- new names may be added at every time,

- the user may use names not contained in the database.

In Datenbank-DIALOG the following technique is used. Proper names are not listed in the lexicon. Whenever Datenbank-DIALOG encounters an unknown word it takes it as a proper name. Only if that hypothesis fails for grammatical reasons, it goes into spelling correction.

There are some problems with this solution. Some proper names are also words of the language (e.g. Black, Smith). Such proper names must be explicitly entered into the lexicon both as a regular word (adjective, noun, etc.) and as a proper name. This of course may lead to consistency problems again.

Proper names may be compounds (e.g., United Airlines) and they may inflect (e.g., Miller's, the Johnsons).

5.3 Meta knowledge

There are to different types of knowledge I would call metaknowledge in the course of database queries in natural languages. One concerns knowledge about the dialog itself:

as, e.g., the query *What do you mean by saying that?*

It seems to be impossible to provide a practical NLI with that kind of knowledge. Instead, one must clearly tell the user that he must not expect such behaviour.

The second type of metaknowledge is knowledge about ones own knowledge. In other words an NLI must know what it knows about. It should have at least a minimal explanation facility. A simple way is to allow questions to the data dictionary of the data base. This enables simple questions about the contents of the data base. The quality of that approach clearly depends on the quality of the data dictionary.

6 Robustness

Like for all software systems robustness is of utmost importance for NLIs. A robust NLI should:

- make it easy to create new applications, to change and fix existing ones,

- behave in a co-operative and natural way (this includes habitability), and

- respond intelligently to user failures.

6.1 Maintaining and changing the data

Acceptability of an NLI depends not only on the satisfaction of the end user but also on that of the systems administrator who is responsible to create and service applications. Accordingly, it is crucial for an NLI that the necessary data structures can be manipulated as easy as possible. To achieve this goal one needs both an adequate system architecture and efficient tools for data acquisition.

6.1.1 Keeping syntax and semantics separate

In the field of NLIs there is a long tradition of syntagmatic systems (e.g., Hendrix & Walter 1987) and systems with no clear boundary between syntax and semantics (e.g., the FRUMP system, DeJong 1982). While these systems may exhibit a good performance they have some serious drawbacks. Because of the lack of syntactic generalizations their grammar has to be re-designed for every new application. This is clearly in contrast to requirements on portability.

Accordingly, state-of-the art systems exhibit a distinctive syntactic component (e.g., Bates, Moser & Stallard 1984, Ginsparg, 1983, Grosz et al. 1987, Thompson & Thompson 1985). In *Datenbank-DIALOG* syntax and semantics are represented in different data structures, an annotated context free grammar and case frames respectively. Processing, on the other hand, works in parallel (Trost et al. 1988).

6.1.2 Core lexicon

The lexicon is at the heart of every NLI. It contains data for morphology, syntax and semantics/pragmatics. Although the lexicon must be tailored to every new application a considerable part of the data will stay the same. This part which we call core lexicon must be provided together with the system.

It clearly depends on the architecture of the NLI how much data can be included in the core lexicon. A clear distinction between structure (morphology and syntax) and content (semantics and pragmatics) allows for more data to be contained in the core lexicon.

What will be contained in an appropriate core lexicon:

- all function words (i.e. articles, conjunctions, auxiliaries),

- words from closed word classes (i.e. prepositions, pronouns),
  but: Their semantics may be domain-dependent.

- morphologically irregular words (i.e. verbs following strong conjugation),
  but: Their syntax and semantics are domain-dependent.
  Morphologically, they may be used in derivation to produce new words.

As we have seen, the core lexicon may contain words which are only partially defined. These words must be marked so that they are not used in the parsing process. It is still useful to include them into the core lexicon. The inclusion of irregular words makes the morphological classification of newly defined words much easier: the inclusion of prepositions eliminates the need to ask for their syntactic restrictions (like case) and they may be used for prepositional objects without asking further information (like *at in look at*); and so on.

6.1.3 Tools for data acquisition

For the user data are centered around words. Tools should facilitate the insertion of new words considering:

- Morphology: To which morphological paradigm does the word belong.

- Syntax: To which syntactic class does the word belong.

- Semantics: Word meaning must be related to existing semantic predicates (or new ones must be established). The meaning must be integrated into the linguistic model.

- Pragmatics: The connection between semantic predicates and database relations must be established.

The syntax component is responsible for structural generalizations, e.g. active - passive, attributive use of verbs. As a result, once a word (with its meanings) is entered it can be used in all the usual ways supported by the NLI.

The ease of acquiring new words depends on the size of the core lexicon, the clear distinction between syntax and semantics on the one hand and linguistic model and database model on the other hand. Of course it also depends on the quality of the acquisition tool (see, e.g., Ballard 1986 and Ballard & Stumberber 1986).

6.2 Robustness - reacting sensibly

In this chapter we will deal with all the aspects of adapting to the user's special needs. This could be requests to interpret some input in a specific way or to know how some data are represented in the database contrary to the user's expectations.

6.2.1 Patterns

In most domains there exist conventions on how to represent certain facts. Abbreviations and special formats are conventionally used in certain domains which are not interpretable in the same way outside of this domain.

e.g., look at these different formats for dates:

29.5.1991, 1991-05-29, 91/29/05

An NLI must provide the possibility to define such patterns in an easy way. Since they have a strict format this usually poses no problem. But there are some things one must keep in mind:

- Interaction between patterns and spelling correction. Usually such patterns cannot be processed if they contain spelling errors.

- Overlapping patterns may cause problems.
  e.g., *Wer hat am 29.5. 1991 DM verdient?*

6.2.3 Coding and decoding

Usually, databases contain a lot of coded data, i.e. data which are not stored in natural language. Let's assume a database where persons have an attribute *sex* stored with them. If this attribute has the values *female* and *male* we can directly use these values. But most probably the encoding will be different:

e.g. *sex* = *f* stands for female and *sex* = *m* for male;

or we may have a different encoding in the form of an attribute female:

e.g. *female* = *yes* stands for female and *female* = *no* for male.

There is also the possibility of having two different tables in the database (e.g. instead of *person* there is *employee* and *client*) having an attribute representing the feature *sex*. In that case we may have one encoding of sex in the first table and a different one in the second.

An even more complicated case arises if a certain feature corresponds to more than one attribute. Let's assume that every table *client* contains the attributes *street, city* and *zip code*. The feature *address* has no direct equivalent in that table then. Instead, we need to combine (concatenate) these three attributes.

Lastly, not only values are encoded. Operators may be encoded as well. Assume an attribute *name* that contains not only the second name but (optionally) also the title of that person (e.g., Prof. Miller). In that case the *EQUAL* operator of the domain model must be encoded as *CONTAINS*.

There must be means to deal with all these phenomena when translating from the language to the data model.

6.2.4 Default output

Finding the right - instead of the correct answer - is one of the most tricky problems in **natural language query** systems. There is a lot of research in areas like over-answering or presuppositions. I will give just one small example for the range of problems encountered in this field.

While in formal query languages like SQL the required output must be explicitly defined, natural language does not force one to do so. Consider the query:

1) Where does Miller live?

This query may yield a wide range of answers. Some examples are the following:

2) In the United States.

3) In New York City.

4) In the Fifth Avenue.

5) 302 Fifth Avenue, New York, 123456 NY, U.S.A.

Each one of these answers may be appropriate depending on the context. There must be a means to specify which of these will be delivered. *Datenbank-DIALOG* requires the definition of a default output for every table of the database. This default output can contain any combination of columns of that table. It may also include information from other tables which has to be joined.

6.3   Robustness - reacting to errors

An important aspect of robustness is how the system can handle failures both of the user and of the system itself.

6.3.1   Paraphrasing the user's question

To deliver a paraphrase of the original user's query serves two purposes: to reassure the user that the NLI is indeed capable to interpret his/her questions correctly, and - in case of a misinterpretation or an unexpected ambiguity - to give the user the opportunity to recognize the failure.

A simple solution is to optionally print out the formal query. Of course, this makes only sense for users who can read such queries. For some user groups this may be meaningful, because it is much easier to read such a formal query than to formulate it. But for the bulk of the applications we envision for NLIs this will not do.

A better solution is to paraphrase that formal query in natural language. But this poses some unexpected problems. We have to balance between exhaustiveness and readability, e.g. long or complex joins (Buchberger & Auterith 1988).

6.3.2   Spelling correction

First, we have to consider what counts as a spelling error. Usually, spelling errors are taken to be typos, i.e. errors occurring due to the use of a keyboard for inputting sentences. Usually spelling correction deals with at most one error per word, which accounts for about 80% of all errors. Errors are:

- omission (exmple),
- insertion (exasmple),
- replacement (exemple), and
- change of position (exampel).

Errors involving the blank character should be regarded as well (exam ple, forexample).

Shorter words have much less redundancy than long ones. Take, e.g. the English words *he, she, me, be, we*. They are separated by just one "error". Correcting these words cannot be done unless the system makes use of the context. Current spelling correction systems usually do not take into account the syntactic and semantic context nor do they include ergonomic considerations like which errors are more plausible to occur because of the use of a keyboard (Dorffner & Trost 1985).

Besides of getting too many corrections for a misspelled word there are some subtler problems: A word which can be recognized correctly may actually he (= be) a misspelling.

Usually, misspelled words are compared against the lexicon. In case of a full-form lexicon this match is simple. But in the case of a lexicon containing only canonical forms (lexemes) the match is more complicated, e.g., *exemples* must first be separated into *exemple* +*s* and then matched against example. More advanced morphological components (e.g. two-level morphology) may necessitate even more integration work.

If proper names are not contained in the lexicon this poses some additional problems for spelling correction:

- Proper names may be wrongly interpreted as misspellings.

- Misspellings may be wrongly interpreted as proper names.

- There may be misspellings in proper names.

Like proper names, formatted data (patterns) cannot be subject to spelling correction. Lastly, unknown words cannot be distinguished from misspelled words. One way out could be the use of a morphological component based on morphs. Such a component may be able to at least recognize unknown words as such.

6.3.3 Error handling

Every NLI will sometimes be unable to interpret a question correctly. In that situation the NLI should give the user some hint why it has failed to understand:

- If a word was not recognized: Inform the user about the unknown word and try to interactively repair the utterance. This is standard technology and is usually done together with spelling correction.

- If the problem is in the syntax or the sentence semantics the user would like to know which part of the question caused the problem. But usually parsers cannot tell the reason behind the failure, but just where they stopped trying. This is very annoying for the user who would of course like to know why her/his query did not succeed. More robust parsers may bring a solution.

7   Conclusion

In the current literature NLIs have been judged mainly according to their linguistic coverage. Very few serious evaluation studies have been performed. This is due to the small group of actual users of NLIs which makes field studies almost impossible. In experimental settings it is often unclear what to test and how to compare results. Nevertheless, looking at the results and our own experiences with an NLI a pattern can be identified.

There are some serious shortcomings in linguistic coverage, mainly in the fields of anaphora and determiners. Instead of waiting for general solutions designers of NLIs should try to focus on solutions in the restricted context of database querying.

The acceptability of an NLI depends at least as much on portability, habitability and robustness as on linguistic coverage. Accordingly, more effort is required to solve these problems adequately if NLIs are to fulfil the expectations placed on both their usefulness and their wide circulation.

Ad-hoc solutions clearly have their limitations. Adding features means adding unexpected side-effects. What is important is to create a stable and predictable system behaviour. Sometimes simpler techniques are preferable to more advanced ones because they are easier to integrate.

8   Acknowledgements

Financial support for the Austrian Research Institute for Artificial Intelligence is provided by the Austrian Federal Ministry for Science and Research. I would like to thank Ernst Buchberger, Hans Haugeneder and Hans Kamp for helpful comments on this paper. I also want to thank Professor Robert Trappl for his continuing support.

9   References

[1]   Ballard B. (1986): User Specification of Syntactic Case Frames in TELI, A Transportable, User-Customized Natural Language Processor; Proceedings of the 11th International Conference on Computational Linguistics, Bonn, pp.454-460.

[2]   Ballard B., Lusth J., Tinkham N. (1984): LDC-1: A Transportable Natural Language Processor for Office Environments; *ACM Transactions on Office Information Systems* 2(1)1-23.

[3]   Ballard B., Stumberber D. (1986): Semantic Acquisition in TELI: A Transportable, User-Customized Natural Language Processor; Proc. 24th Annual Meeting of the ACL, Columbia University.

[4]   Bates M., Bobrow R.J. (1984): Natural Language Interfaces: What's Here, What's Coming, and Who Needs It; in W.Reitman (ed.), Artificial Intelligence Applications for Business, Ablex, Norwood, NJ.

[5]   Bates M., Moser M., Stallard D. (1984): The IRUS Transportable Natural Language Interface; Proc. First International Workshop on Expert Database Systems, Kiawah Island, pp.258-274.

[6]   Berwick R.C.: Intelligent Natural Language Processing (1987): Current Trends and Future Prospects; in W.E.Grimson, R.S.Patil (eds.), AI in the 1980s and Beyond, MIT Press, Cambridge, MA.

[7]   Buchberger E., Auterith A. (1988): Generating German paraphrases from SQL expressions; in R.Trappl (ed.), Cybernetics and Systems '88, Kluwer, Dordrecht, 885-892.

[8]   Damerau F.J. (1964): A Technique for Computer Detection and Correction of Spelling Errors; *Comm. ACM* 7(1964),171-176.

[9]   DeJong G. (1982): An overview of the FRUMP system; in W.G.Lehnert, M.H.Ringle (eds.), Strategies for Natural Language Processing, Lawrence Erlbaum, Hillsdale, NJ.

[10] Dorffner G., Trost H. (1985): Morphologische Analyse und intelligente Fehlerkorrektur in VIE-LANG; H.Trost, J.Retti (eds.), Österreichische Artificial Intelligence-Tagung 1985, pp.42-53, Springer, Berlin.

[11] Ginsparg J. (1983): A Robust Portable Natural Language Database Interface; Proc. Conference on Applied Natural Language Processing, Santa Monica, CA, pp.25-30.

[12] Grosz B.J., Appelt D.E., Martin P.A., Pereira F.C.N. (1987): TEAM: An Experiment in the Design of Transportable Natural-Language Interfaces, *Artificial Intelligence* 32(2)173-244.

[13] Harris L.R. (1984): Natural Language Front-Ends; in P.H.Winston, K.A.Prendergast (eds.), The AI Business, MIT Press, Cambridge, MA.

[14] Hendler J.A., Michaelis P.R. (1983): The Effects of Limited Grammar on Interactive Natural Language; Proc. of CHI'83: Human Factors in Computing Systems, pp.190-192, ACM, New York.

[15] Hendrix G.G., Walter B.A. (1987): The Intelligent Assistant - Technical Considerations Involved in Designing Q&A's Natural-Language Interface; *BYTE* 12(14)251-258.

[16] Hill I.D. (1983): Natural Language Versus Computer Language; in M.E.Sime, M.J.Coombs (eds.), Designing for Human-Computer Communication, Academic Press, London.

[17] Jarke M., Turner J.A., Stohr E.A., Vassiliou Y., White N.H., Michielsen K. (1985): A Field Evaluation of Natural Language for Data Retrieval, *IEEE Transactions on Software Engineering*, SE-11(1).

[18] Johnson T. (1985): Natural Language Computing: The Commercial Applications; Ovum, London.

[19] Keenan E.L., Stavi J. (1986): A Semantic Characterization of Natural Language Determiners, *Linguistics and Philosophy* 9, 253-326.

[20] Krause J. (1982): Mensch-Maschine-Interaktion in natuerlicher Sprache, Niemeyer, Tuebingen.

[21] Ogden W.C., Sorknes A. (1987): What Do Users Say to Their Natural Language Interface?; in H.-J.Bullinger, B.Shackel (eds.), Human-Computer Interaction - INTERACT '87, North-Holland, Amsterdam.

[22] Rettig M., Bates M. (1988): How to Choose Natural Language Software; *AI-Expert* 3(7)41-49.

[23] Saebo K.J. (1988): A Cooperative Yes-No Query System Featuring Discourse Particles, Proceedings of the 12th International Conference on Computational Linguistics, Budapest, pp.549-554.

[24] Schur S. (1988): The Intelligent Database; *AI-Expert* 3(1)26-34.

[25] Tennant H.R. (1980): Evaluation of Natural Language Processors; Report T-103, Coordinated Science Laboratory, University of Illinois - Urbana.

[26] Thompson B., Thompson F. (1985): ASK is Transportable in Half a Dozen Ways; *ACM Transactions on Office Information Systems* 3(2)185-203.

[27] Trost H., Buchberger E., Heinz W., Hörtnagl C., Matiasek J. (1987): "Datenbank-DIALOG" - A German Language Interface for Relational Databases; *Applied Artificial Intelligence*, 2(1)181-203.

[28] Trost H., Buchberger E., Heinz W. (1988): On the Interaction of Syntax and Semantics in a Syntactically Guided Caseframe Parser; Proceedings of the 12th International Conference on Computational Linguistics, Budapest, pp.677-682.

The Semantics Application Interface

Joachim Laubsch
Hewlett-Packard Laboratories, 1AU
1501 Page Mill Road, Palo Alto, CA 94304-1126
laubsch@hplabs.hpl.hp.com

1   Introduction

Mapping natural language into a formal language with well defined semantics is prerequisite but not enough for practical applications such as "front ends" to database systems or other applications. We have to translate the "logical form" produced by the semantics module into a formal language understood by the application. Many applications allow a programmatic interface, and some applications (such as databases) define an interface language, e.g. **SQL** which has become a standardised language. For communication with knowledge bases, a standard communication language — **KIF** "Knowledge Interchange Format" — is emerging ([7]), making it possible to query or update knowledge bases written in different representation languages.

The logical language for representing the meaning of natural language expressions — *NLL* — was designed as part of Hewlett Packard's natural language system (HP-NL [8]) to[1]

- facilitate construction of the semantic structures from the syntactic and lexical structures

- provide an interface language for various transformations that lead from the initial semantics to an application specific semantics. Such transformations include:

  NLL-**specific inferences** i.e. inferences which are valid independently of the application domain.

  Simplification i.e. meaning preserving transformations that produce a canonical form.

---

[1]   *NLL* reflects common practice in computational linguistics (see [1]) and is described in [5].

Disambiguation e.g. translation from discourse representation theory (DRT) to the logic of generalized quantifiers.

Domain specific inferences

Translation to an application language e.g. mapping *NLL* forms into SQL programs or the NewWave task language.[2]

The next section gives a brief overview of *NLL*. Sections 3 to 4.1 illustrate by example the above mentioned transformations.

2   A brief overview of *NLL*

In many respects *NLL* is similar to the representation language developed within the "core language engine" (CLE [1]) at SRI-Cambridge. *NLL* is based on the language of generalized quantifiers [2] and has the following features:

- *Arbitrary arity (anadicity of)* predicates, e.g. the following are atomic formulas:

$$\text{Manage(agt:Jones pat:Smith)}$$
$$\text{Manage(agt:Jones project:NewWave)}$$

  An argument is marked by a keyword (agt, pat, project) followed by ":". Predicates and functions take a set of such keyworded arguments (without duplicate keywords).

- *Generalized quantifiers:* The notion of a generalized quantifier[3] is based on the relation between two sets — a restricton set R and a scope set B. $R \cap B$ is called the intersection set I. Within *NLL*, a generalized quantifier $\mathbf{Q}$ is defined as a binary predicate:[4]

$$\lambda(n,m)\mathbf{Q}(n,m)$$

  which is applied to the cardinality of the restriction set, $|R|$, and the cardinality of the intersection set $|R \cap B|$. This is the number-theoretic characterization of generalized quantifiers. There is an equivalent

---

[2] The NewWave Task language is a programming language, developed at HP, where relations, actions and objects are confined to the domain of a desktop menu/mouse/windows user interface.

[3] We include only natural language quantifiers which are "quantificational", excluding possessives, vague quantifiers, and quantifiers expressing defaults (such as "usually").

[4] For a comprehensive introduction and survey of quantifiers in logic and linguistics see [10].

formulation in terms of sets and relations among sets (for a comparison see [10]).

The syntax of a quantified formula is:

$$(\langle \text{quantifier} \rangle \; \langle \text{var} \rangle \; \langle \text{restriction set} \rangle \; \langle \text{body set} \rangle)$$

Example: (Most ?z Man(inst:?z) Sleep(inst:?z))

has the interpretation: "The number of men who sleep is more than half of the number of men." In the set-theoretic interpretation this is equivalent to:

$$Q_{most}(\overbrace{|\{x|\text{Man}(x)\}|}^{\text{restriction}}, \overbrace{|\{x|\text{Man}(x)\}\cap\{x|\text{Sleep}(x)\}|}^{\text{intersection}})$$

where

$$Q_{most} =_{def} \lambda(r,i)\, i > r/2$$

That is

(most ?x ϕ(?x) ψ(?x)) is true iff $|R - I| > |I|$

Among the generalized quantifiers are $\forall$, $\exists$, "Most", "The", "1", "2", ... as well as complex determiners such as:

$((\{= 1\})$?x ϕ(?x) ψ(?x)) is true iff $|I| = 1$

...

$((\{= ?n\})$?x ϕ(?x) ψ(?x)) is true iff $|I| = n$

Similarly for the specifiers $<, \leq, \neq, >, \geq$

Example:

"More than five men smoke."

$((\{> 5\})$?x man(inst:?x) smoke(agent:?x))

This is true iff the intersection of the set of men and the set of objects that smoke is greater than 5.

Instead of a constant after the specifier, there may be a variable following. This is useful in

Comparatives: "More consultants work than engineers teach."

> (exists ?n NatNum(inst:?n)
>> and{({(= ?n)} ?m Engineer(inst:?m) teach(agt:?m))
>> ({(> ?n)} ?c Consultant(inst:?c) work(agt:?c))})

"How many" questions: "How many men smoke?"

> (?lambda ?m Measure(inst:?m)
>> (({= ?m}) ?x man(inst:?x) smoke(agent:?x)))

- *Question Wff*

  $\langle$Question Wff$\rangle$ ::= ($\langle$Questioner$\rangle$ $\langle$Scope$\rangle$)

  $\langle$Questioner$\rangle$ ::= ?lambda $\langle$Variable$\rangle$, ... $\langle$Restriction$\rangle$

  $\langle$Scope$\rangle$ ::= $\langle$Wff$\rangle$

  $\langle$Restriction$\rangle$ ::= $\langle$Wff$\rangle$

The meaning of (?lambda ?x ϕ(?x) ψ(?x)) is a predicate characterizing the set of all true propositions that arise from substituting ?x by some denoting term in

$$\text{and } \{\phi(?x)\psi(?x)\}$$

Examples:

> "What did O'Brian sell?"
> (?lambda ?x Thing(inst:?x) sale(agent:O'Brian product:?x))

> "Which woman manages what department?"
> (?lambda ?x,?y
>> and{Woman(inst:?x) Department(inst:?y)}
>> manage(agt:?x pat:?y))

- *Restricted Parameters (Indeterminates):*

$\langle$Restricted Parameter$\rangle$ ::= ([˙$\langle$Determiner$\rangle$˙] ?$\langle$Identifier$\rangle$ | $\langle$Restriction$\rangle$)
$\langle$Restriction$\rangle$ ::= $\langle$Wff$\rangle$

A restricted parameter[5] restrains a variable to an object of the domain for which the restricting formula holds. Any occurrence of the variable within

---

[5] Notation: We use BNF syntax with non-terminals enclosed in "$\langle\rangle$" brackets. Optional constituents are enclosed in "[˙ ˙]" brackets. Alternation is indicated by "˙|". This way we are free to use "|, [,]" as terminals of *NLL*.

some *NLL* expression that would otherwise be considered free — as a consequence of introducing the restricted parameter — becomes a restrained occurrence. Therefore, a restricted parameter allows the representation of scope ambiguity.

There may be more than one restricted parameter restraining the same variable. The optional determiner of a restricted parameter specifies the quantificational force of the restraint on the variable. The default quantificational force is the existential-determiner, but any determiner — like in a quantified Wff — can be used.

Example:

"Jones hired a secretary."
hire(agent:Jones patient:(?w I Secretary(inst:?w)))

"Jones hired two secretaries."
hire(agent:Jones
    patient:(({= 2}) ?w I Secretary(inst:?w)))

3   *NLL* and its inferences

3.1   *NLL* simplifications

3.1.1 Implied Conjunct Removal

First we consider a general rule.

In a conjunction of atoms p

$$\wedge \{p_1 \dots p_i \dots p_n\}$$

p_i can be eliminated if p_i can be derived from the other conjuncts, that is if:

$$(1) \qquad \{p_1 \dots p_n\} \setminus \{p_i\} + p_i$$

A special case of this rule is the following:

Subsumption of a Literal by Specialization   Atomic conjunct p *subsumes* atomic conjunct q if they are both atoms with the same predicate and

$$(2) \qquad \text{Arg-set}(p) \subseteq \text{Arg-set}(q)$$

where Arg-set is the function that returns the arguments of an atomic wff. The specialized conjunct q implies the more general conjunct p which will be removed from the conjunction.

Example   "Which areas are in the Midwest region?"

```
(?lambda ?z area-name(inst:?z)
   (the ?x =(?x,MidWest)
     (exists ?y
       and{area-name(inst:?z region:?y))
         region(inst:?y name:?x)}))
```

We apply the rule *Move-quantifier-up*[6] (The quantifier exists is moved up into the scope of the quantifier the.) This leads to ($\rightarrow$):

```
(?lambda ?z area-name(inst:?z)
   (the ?x
     (exists ?y and{=(arg1:?x arg2:MidWest)
               area-name(inst:?z region:?y)
               region(inst:?y name:?x)}))))
```

Identity-substitution for ?x with elimination of the now vacuous quantifier (the ?x ...).$\rightarrow$

```
(?lambda ?z area-name(inst:?z)
      (exists ?y
          and{area-name(inst:?z region:?y)
              region(inst:?y name:MidWest)}))
```

Move-quantifier-up (The quantifier exists can now be moved into the restrictor of the questioner ?lambda) :

```
      →

(?lambda ?z (exists ?y
        and{area-name(inst:?z)
            area-name(inst:?z region:?y)
            region(inst:?y name:MidWest)}))
```

Finally, we can apply the Subsumption-by-Specialization rule (2) to the restrictor of the exists form. The first conjunct is eliminated since its argument set is a subset of the second conjunct's argument set.

```
      (?lambda ?z (exists ?y
           and{area-name(inst:?z region:?y)
               region(inst:?y name:MidWest)}))
```

---

[6]   This is an example of a "normalization rule", a trivial rule on which the applicablility of simplification rules depends.

All rules used in the previous example follow from the semantics of *NLL*, hence are reusable in extensions of *NLL* and applications to arbitrary domains.

Domain specific rules   For a particular application we axiomatize a domain theory, *A*. Part of *A* may be a theory of sorts, as for instance expressed in a sortal hierarchy. Then we may use the derivability relation of that theory $\vdash_A$ and equation (1) to eliminate implied conjuncts.

Rewrite Rules for Disambiguation   An *NLL* formula may be semantically ill-formed. We make use of heuristic rules to rewrite such formulas. Violation of sortal constraints is the prime reason of ill-formedness. The syntax of such rewrite rules is:

⟨Condition⟩ : ⟨Wff⟩ → ⟨Wff⟩

The following example shows how the location role of the predicate located can be coerced to the location of the object filling that role:

```
Type-of(subtype:?y type:Location) :
located(thm:?x location:?y) →
  and{located(thm:?x location:?z)
      located(thm:?y location:?z)}
```

Bridge Axioms   In order to express the meaning of an *NLL* formula in terms of the predicates and relations of an application domain we use bridge axioms that translate between theories. The notation for a bridge axiom uses ↦, the relation between *NLL* formulas of two theories (which should not be confused with logical implication):

⟨Condition⟩ : ⟨Wff⟩ ↦ ⟨Wff⟩

Example:

```
Type-of(subtype:?y type:Location) :
located(thm:?x location:?y) ↦
    and{area-name(inst:?x region:?y)
        region(inst:?y name:?x)})
```

Here the English predicate "located" is mapped into application specific predicates "area-name" and "region" which were defined in the theory derivative of the application's database scheme.

We found that often bridge axioms and rewrite rules for disambiguation introduce extra conjuncts which can later be eliminated by *NLL* simplifications. The benefit of these core *NLL* simplifications is that they can be reused at various stages of processing the semantic form.

3.1.2 Inferences for Group and Location Terms

Group Terms $\langle$Group Term$\rangle$ ::= +{$\langle$Term$\rangle$, $\langle$Term$\rangle$}

The group constructor +{ } is associative and commutative. We also use the abbreviated notation for the composition of n (n > 2) groups:

+{$g_1 g_2$, ... g_n} $\equiv$ +{g_1, +{g_2 ... g_n}}

Example: +{T, D, H} $\equiv$ +{T, +{D, H}}

Group terms can be interpreted either *distributively or collectively*. An example where a distributive inference can be made is:

work(agent/i:+{John, Mary})
$\rightarrow$ and{work(agent:John) work(agent:Mary)}

Group terms may have *sigma* terms as components. A sigma term designates a group of individuals — the sum of all individuals for which a restriction holds. Sigma is a variable binding operator which restrains its variable like a restricted parameter does. Its syntax is:

$\langle$Sigma Term$\rangle$ ::= (sigma $\langle$Variable$\rangle$ | $\langle$Restrictor$\rangle$)

A sigma term denotes a particular *individual sum* defined (by Link [6]) as

$$\sigma x P(x) = \iota x(*P(x) \wedge \forall_y (*P(y) \rightarrow y \leq x))$$

where * is an operator on 1-place predicates P which generates all the individual sums of members of the extension of P. Then (sigma ?x P(?x)) denotes the supremum of *P(x).

Example: "Mary and the men work."

work(agent/i:+{Mary,(sigma ?m | Man(inst:?m))})
$\rightarrow$ and {work(agent/i:Mary)
            (forall ?m Man(inst:?m) work(agent/i:?m))}

Location Terms   A similar inference rule holds for location terms, which have the syntax:

$\langle$Location Term ::= reg-X{$\langle$Term$\rangle$, ...}

reg-X ("regional intersection") is a function whose domain and range are locations. Similarly, there are unary functions like "in", "at", and "on" ranging over locations. Function terms constructed from these are transformed into corresponding roles:

P(inst:?x location:reg-X{in(SJO), at(IBM)})

$\rightarrow$ P(inst:?x in:SJO at:IBM)

Repeated occurences of the same function in a regional intersection are transformed into a conjunction:

P(inst:?x location:reg-X{in(SJO), at(IBM), in(CA)})
$\rightarrow$ and{P(inst:?x at:IBM in:CA) P(inst:?x at:IBM in:SJO)}

3.2 From DRT to LGQ

DISCOURSE REPRESENTATION THEORY (DRT) recommends nonlinear representations — Discourse Representations — for treatment of anaphora and quantification (cf. [4], [3]). The language of SITUATION THEORY attempts to analyze the same material using RESTRICTED PARAMETERS, which appears to be a much milder extension of the logical language. We will now describe the steps of transformation from an initial semantics *NLL* form containing such restricted parameters into a form that does not contain them and which is purely in the language of generalized quantifiers.

Predetermined Scope: Some forms already have a predetermined scope. In the following example the ?y is given wide scope (at the negation) during syntactic analysis[7]:

```
"The consultant was not hired by a manager."
~hire(agent:(?x | Manager(instance:?x))
        patient:(the ?y | Consultant(instance:?y)))

→
(the ?y Consultant(instance:?y)
   ~(exists ?x Manager(instance:?x)
           hire(agent:?x patient:?y)))
```

A quantified wff whose determiner is the quantificational force of the restricted parameter and whose scope is the transformed formula of the entire form (with the restricted parameter replaced by its variable) is constructed. We say that the restricted parameter (the ?y | Consultant(instance:?y)) is *expanded at* ~hire(...).

---

[7] Scopal disambiguation of a restricted parameter (by syntactic analysis or otherwise) cannot be expressed in the concrete syntax of *NLL*. But the abstract syntax provides a convention for adding arbitrary attribute/value pairs to any node in the abstract syntax tree.

Default Scoping: The first step is to find the place in the abstract syntax tree of a formula, where a restricted parameter should have its default scope.

- Find the narrowest scope for each restricted parameter that would bind all its occurences.

- Propagate a restricted parameter to its ancestor according to the following constraints (ordered by their priority)

  1. No restricted parameter may have a larger scope than a restricted parameter it depends on (occuring free within its restrictor).
  2. A restricted parameter will not propagate up to quantified, negated, or conditional formulas.
  3. By default, a restricted parameter is propagated up except at atomic formulas.

The above example also shows how the restricted parameter for ?x is propagated up to the atomic wff hire(agent:?x) but not up to the negation. Since a restricted parameter by default has existential quantificational force, exists is wrapped around the atomic wff.

Ordering restricted parameters: At this stage, a node of the abstract syntax tree of the formula may have a set of restricted parameters attached. We could order this set, if we had a strict order relation, but the "depends on" relation is not antisymmetric. That means that there will be situations with mutually dependent restricted parameters. An example is the Bach/Peters sentence

```
"The Pilot who shot at it hit the MIG that fired at him."
hit(agt:(?p | and{Pilot(inst:?p) shoot-at(agt:?p obj:?m)})
       obj:(?m | and{MIG(inst:?m) fire-at(agt:?m obj:?p)}))
```

Fortunately, in this interpretation both ?p and ?m have existential force, and *NLL* allows sequences of bound variables in existential and universal quantifiers:

```
   →
(exists ?p,?m and{Pilot(instance:?p)
                  shoot-at(agent:?p obj:?m)
                  Mig(instance:?m)
                  fire-at(agent:?m obj:?p)}
            hit(agent:?p obj:?m))
```

If we exclude mutually dependent restricted parameters, we can define the "depends on" relation as a strict partial order. (For antisymmetry to hold in this relation, those restricted parameters which do not depend on any parameter, are ordered lexicographically.) At each node we then form a sequence of restricted parameters that does not violate the order relation. This sequence determines the nesting of the quantified expressions to be built that guarantees that all occurrences of a restricted parameter's variable is bound.

Wrapping restricted parameters: We process the abstract syntax tree of the formula top down, and at a node which has a nonempty sequence of restricted parameters we perform the following transformation:

1. At a quantified Wff if the restricted parameter occurs in both quantifier and scope (so called "donkey" sentence)
   - In case of an existential or universal quantifier, add the variable to the bound variables and the restriction of the restricted parameter to the restriction of the quantifier.

   Example:

   ```
   (forall ?x and{Farmer(inst:?x)
                       own(agt:?x th:(?z|Donkey(inst:?z)))}
             beat(agt:?x th:?z))
   ```

   ```
   (forall =?x,?z and{Farmer(inst:?x)
                       own(agt:?x theme:?z)
                       Donkey(inst:?z)}
             beat(agent:?x theme:?z))
   ```

   - For all other quantifiers, if the restricted parameter occurs in the quantifier, replace the literal[8] that contains it by its expansion (s.a.). Add that literal to the scope and expand the restricted parameter at the scope.

   Example:

   ```
   (most ?x and{Farmer(inst:?x)
                    own(agt:?x th:(?z|donkey(inst:?z)))}
           beat(agt:?x th:?z))
   ```

   ```
   (most ?x and{Farmer(inst:?x)
                    (exists ?z Donkey(inst:?z)
                               own(agt:?x theme:?z))}
         (exists ?z1 Donkey(instance:?z)
                     and{own(agent:?x theme:?z1)
                         beat(agent:?x theme:?z)}))
   ```

2. At all other nodes expand the restricted parameter at the node.

---

[8] A literal is an atomic wff or a negated atomic wff.

4   Programming the application interface

4.1   Transformations for Relational Data Bases

Inferences within *NLL***:** Data base models use the concept of a "key", i.e. an attribute (or a set of attributes) whose values uniquely identify each entity. There are several inferences that can be drawn from information about *key-roles*, and we mention one of them here:

In NLL_{rdb}, if *Key-Role-Value*($\bar{x}$) = Key-Role-Value($\bar{y}$) then

$$P(\bar{x}) \wedge P(\bar{y}) \equiv P(\bar{z})$$

where

$$z = \{ r \mid (r \in \bar{x} \vee r \in \bar{y}) \wedge \neg\text{Key-role}(r,P) \}$$

Example:

```
and{Orders(engineer:Jones sales:?x KEY:?k)
    Orders(customer:007 KEY:?k) }
```

$\equiv$

```
    Orders(customer:007 engineer:Jones sales:?x)
```

4.2   Mapping *NLL* to SQL

A fundamental equivalence between predicate calculus and relational calculus (of which SQL is a subset) is the following:

$$(3) \qquad\qquad P(\bar{x}) \Leftrightarrow \exists_{p \varepsilon P} \bigwedge_{i=1}^{n} p.i = x_i$$

where $\bar{x}$ stands for $1:x_1, 2:x_2,...n:x_n$, and P is the relation corresponding to the predicate P. In other words, P holds of some objects $\bar{x}$ iff the relation P contains a tuple p whose fields are equal to the corresponding objects. In *NLL* this equivalence is slightly more complicated, since the keys are symbols rather than natural numbers. Therefore, the above equivalence can be applied after $P\bar{x}$ has been translated into an application specific extension of *NLL* whose predicate vocabulary contains just the primitive application specific predicates (where predicates correspond to tables and roles to fields).

In order to express (3) we need a new language that contains relations, tuple variables, field accessors etc. One idea is to *extend* the definition of *NLL* by these constructs and then express (3) as a *transformation rule* between these two languages (using the pattern matching syntax of REFINE):

```
'@P(@rl:@x1 .. @rn:@xn)'
--> '(exists (@p in @P)
      and{ @p.@rl = @x1 .. @p.@rn = @xn})'
```

This transformation rule consists of an antecedent and a consequent (separated by "-->"). In the antecedent we have a *NLL* formula containing meta-variables (preceded by "@"). In the consequent we have a formula in extended *NLL*, mentioning the same meta-variables. The antecedent specifies a decomposition pattern, and the consequent a composition pattern. The patterns are shorthand notation for abstract syntax trees.

We used Refine (see [9]) to define languages and specify transformation rules. The Refine compiler would translate these transformation rules to Common Lisp functions.

4.3   Example of a transduction to SQL

As an example of how a transformation rule is stated as a Refine program, we consider the rule that transforms an extended *NLL* existential quantifier into an SQL SELECT statement:

```
rule existential-to-SQL-select (a)
  a = 'exists ($vs)[@pred]'
  -->
  a = '0 < select count(*)
        from $(make-aliases(vs))
        where @pred'
```

The function make-aliases (not given here) constructs from a list of predicates a list of names of tables and tuple-variables. The transduction rules would for instance generate the SQL program:

```
0 < (select COUNT (*)
      from SUPERSORTS_TRC SUPERSORTS_TRC_1,
          ARTWORKS ARTWORKS_2, ART_MEDIA ART_MEDIA_3
      where ARTWORKS_2.ARTWORK = 'F0612' AND
          ARTWORKS_2.ARTIST = 'VANGOGH' AND
          SUPERSORTS_TRC_1.SUPERSORT$ = 'PAINTMED' AND
          ART_MEDIA_3.ARTWORK = 'F0612' AND
          ART_MEDIA_3.MEDIUM = SUPERSORTS_TRC_1.SORT)
```

from the *NLL* form (representing "Van Gogh painted Starry Night."):

```
(exists ?x supersorts_TRC(SORT:?x SUPERSORT$:PaintMed)
    and { artworks(ARTIST:VanGogh ARTWORK:F0612)
        art_media(MEDIUM:?x ARTWORK:F0612) })
```

The fundamental equivalence (3) was appplied three times (to the predicates supersorts_TRC, artworks, and art_media). The scope of the exists form was moved into the restrictor. Finally, the rule existential-to-sql-select (s.a.) was applied.

These transformation rules allow a separation of the meaning representation from the target application. We can easily substitute these rules by rules tuned to a different database query language, or — more generally — to the KIF knowledge interchange language for which presumably there would exist translators to numerous knowledge bases.

5 References

[1] Alshawi et al. Research programme in natural language processing. Final report, Alvey Project No. ALV/PRJ/IKBS/105, SRI Cambridge Research Centre, July 1989.

[2] Jon Barwise and Robin Cooper. Generalized quantifiers and natural language. Linguistics and Philosophy, 4, 1981.

[3] Michael R. Genesereth, Richard Fikes, et al. Knowledge Interchange Format, Version 3.0. Reference Manual. Report Logic-92-1, Logic Group Report, Computer Science Department, Stanford University, Stanford, June 1992.

[4] Irene R. Heim. *The Semantics of Definite and Indefinite Noun Phrases*. PhD thesis, University of Massachusetts at Amherst, 1982.

[5] Hans Kamp. A theory of truth and semantic representation. In J.Groenendijk et al., editor, *Formal Methods in the Study of Language*. Mathematical Centre, Amsterdam, 1981.

[6] Joachim Laubsch and John Nerbonne. An Overview of *NLL*. Technical report, Hewlett-Packard Labs, 1991.

[7] Godehard Link. The logical analysis of plurals and mass terms: A lattice-theoretical approach. In Rainer Bäuerle, Urs Egli, and Arnim von Stechow, editors, *Meaning, Use, and the Interpretation of Language*, pages 302-323, Berlin, 1983. de Gruyter.

[8] John Nerbonne and Derek Proudian. The HP-NL system. Technical report, Hewlett-Packard Labs, 1987.

[9] Douglas R. Smith, Gordon B. Kotik, and Stephen J. Westfold. Research on knowledge-based software environments at KESTREL institute. *IEEE Transactions on Software Engineering*, SE-11:1278-1295, November 1985.

[10] Dag Westerståhl. Quantifiers in formal and natural languages. Report CSLI-86-55, Stanford University, Stanford, 1986. (Also in: D. Gabbay & F. Günthner (eds.) *Handbook of Philosophical Logic*, Vol. IV.).

Verification of Controlled Grammars

Gregor Thurmair
SIETEC Systemtechnik
Carl-Wery-Straße 22, D-81739 München

Abstract

The following paper describes the development of a tool which verifies controlled grammars. First, the development context and the application is described. Then, the definition of the task, and the architecture of the system are given: It consists of a syntactic analyser, a repository of ill-formed structures, a matcher, and an output formulator. Finally, system test results are presented, and some problems to be solved are explained; also, the application environment is sketched.

1   Context

The prototype described in the following paper was developed in the context of the "Translator's Workbench" Esprit Project, based on the METAL linguistic platform. This project aims at improving the documentation and translation process by offering a bundle of software tools, one of which is language checking. This covers spellers, grammar checkers, and controlled grammar verifiers.

The language checking TWB component is the result of trying to optimise the documentation process. Considerations how to improve the input of machine translation were compared with guidelines for technical authors, and large overlaps were detected. This resulted in a common effort to improve both the readability and the translatability of texts, by setting up styleguides for authors. As they define a sublanguage of their own, in that they restrict the grammar of a language they are called controlled languages.

The baseline of the project was the experience that in the case of grammar checking, it turned out that most existing grammar checkers are not reliable

/FRE91/, and therefore are very restricted in their usability. This is due to the fact that most of them do not use a real grammar but are based on some more or less sophisticated pattern matching techniques. However, the fact that they sell shows that there is a need for those tools.

TWB again followed several approaches in grammar and style checking, described in /KUG92/. Among them is a small ATN based NP grammar for German in order to detect agreement errors, which turned out to be the most frequent in German texts (/HEL90/). A second approach has been followed for Spanish grammar checking: Here we used an existing grammar (the METAL analysis) and enriched it by a "peripheral" grammar on top of the core grammar which tries to identify the cases of ungrammaticality (agreement errors, wrong verb argument usage, etc.). During parsing, it can be detected if one of those special rules has fired, and if so, the appropriate diagnostic measure can be taken.

As a result, it turned out that grammar checking needs much more linguistic intelligence if it should be helpful and reliable. It needs fully developed lexicon and syntax components and some "heavy" machinery (in terms of computing power). The TWB tools are the better the more developed the underlying grammars are. However, this hampers their portability to other languages as it means considerable investment.

A last area of language checking was style checking, or better verification of controlled grammars. This is closer related to the documentation business as it tries to implement guidelines for good technical writing, conventions for style and layout, also implying language criteria.

2   Definition

The following paper does not intend to define what "good style" might be; there are many style checkers available (at least for English) which do not do more but formalise their author's intentions. Instead, the goal of the present paper is to present a software tool which allows for the verification of controlled grammars.

Controlled Grammars are subsets of a language, defined for certain purposes and for certain reasons:

- they can be used to support **corporate identity**; e.g. a high-class car producer could ask his authors to use the term "Signalhorn" instead of "Hupe", as "Hupe" sounds cheap.

- they could want to improve the **readability** of a text; this is the main reason to set up guidelines for technical writing. Many companies have this kind of literature; they store what terms and constructions should be used by their authors.

- they want to improve **translatability**; in particular if machine translation is used, restricted text input is required. Examples are Xerox / Systran or Titus (/DUC84/).

- they want to support **non-native speakers'** understanding; this is the basic idea behind the AECMA guidelines or the Caterpillar efforts.

An overview of the development of controlled languages is given in (/ADS90/). The definition of a controlled language is not the focus of this paper, however; the focus is the verification of those languages by software. Verification means to check whether a given document follows such guidelines or not.

A closer look at the existing guidelines shows that there are four kinds of levels on which control must operate:

The first level is **layout control**. Guidelines often contain instructions about the layout, like

- add a line of spaces between two paragraphs,

- use / do not use footers,

- use headers only in certain cases,

- place figures only at the edges of pages.

These layout instructions often determine the choice of the editor; many editors can define their paragraph types accordingly.

A second level is controlled **text presentation**. This refers to the use of abbreviations, the presentation of measure units, and others; examples are

- 5% vs. 5 %

- BRD vs. Bundesrepublik Deutschland

Correct text presentation can to a large extent be controlled by a sophisticated pattern matcher.

A third level of control refers to **controlled terminology**. Often, terminology is defined for a company, and it should be used unambiguously. This is important in particular in large documentation projects with tens of thousands of pages of text. Examples are

- use "Signalhorn" instead of "Hupe"

- use "Diskette" instead of "Floppy"

- use "Handbuch" instead of "Manual"

Control of terminology requires looking up the terms of a text in a Term Bank which stores the status of a term (good, forbidden, unknown, etc.). While term lookup is feasible in large scale applications, it is difficult to verify if a "legal" term is used correctly in a given context.

A fourth level of control is **grammar** instructions. This is the most frequent kind of instructions; examples are

- do not use too long and complex sentences,

- avoid unclear and ambiguous references,

- do not use meaningless words and expressions,

- avoid complex compounds like "Füllgebietsausfüllungsfarbindex" .

Verification on this level requires full syntactic analysis of a given text; it is the most challenging type of controlled grammar verification.

As a result, a controlled grammar verifier must operate on various levels and with various procedures. The following paper concentrates on the most difficult part, however: the verification of grammatical instructions.

3 Implementation

3.1 Architecture

There are two possible architectures for a Controlled Grammar Verifier: Either only the subset described by the Controlled Grammar is implemented; anything that cannot be parsed will be considered to be ill-formed. Or a full parser is implemented and deviations are flagged.

Wenden Sie die Funktion auf die im Beispiel für die Generierung nach der Standardmethode (siehe Kap. 6.2.2.3) beschriebenen Objekte an.

Dieser Satz ist nicht besonders lang,
aber er liegt wie ein Stein im Magen.
Grund dafür ist
das lange Attribut vor "Objekt", nämlich

im Beispiel für die Generierung nach der Standardmethode
(siehe Kap. 6.2.2.3) beschrieben.

Der Satz enthält zu viele Informationen.

Wenn Sie die Informationen auf mehrere Teilsätze verteilen,
können Sie die langen Attribute knacken
und den Satz damit leichter verdaulich machen:

Wenden Sie die Funktion auf die Objekte an, die im Beispiel für die Generierung nach der Standardmethode beschrieben sind (siehe Kap. 6.2.2.3).

Faustregel: Keine langen Attribute!

Übungsbeispiele: ...

FIG. 1: Example of a writing guideline

The former approach is easier to implement but has some drawbacks:

- It is correct that deviations lead to parse failures. But no other parse failures must occur; otherwise a parse failure is not meaningful anymore. This cannot be guaranteed, however.

- No error diagnosis can be given, as a parse failure gives no hints where or why the parser failed. This is considered not to be user-friendly.

We therefore decided to implement the latter strategy. It is more sophisticated in terms of system architecture, but it has the advantage that error diagnosis can be given as a comparison with a "standard" syntactic analysis.

It is an open question whether grammar and style checkers should try to correct the text portions considered to be ill-formed. Our Controlled Grammar Verifier does not intend to do corrections, for two reasons:

- sometimes, stylistic deviations are intended by the authors. In this case, those intentions must be reformulated; this cannot be done by machine at present. Moreover, a system which is too rigid is not acceptable; and it will take some time until sufficient know-how is available to implement correction strategies.

- often, the direction of correction is not known: In case of subject-verb disagreement, for instance, it is not clear whether the subject or the verb should be corrected (if no other hints are available).

Another architectural design topic is the relation between grammar and controlled grammar, i.e. the question how stronlgy integrated the two grammars should be.

Here we decided not to touch the "standard" grammar analysis but to try to do diagnosis on the output of the "standard" grammar. This approach was taken for two reasons:

- It follows the intuition that "illformedness" must be seen on the background of wellformed structures; controlled language is a subset of general language, and the analysis of general language should not be affected by the fact that, later on, a subset has to be identified. Therefore, the controlled grammar verification should operate on "normal" syntactic output.

- From a practical point of view, portability is an issue: It must be taken into account that there are many versions of controlled grammars, focussing on different language issues. However, we do not want to rewrite the grammar every time, as the grammar in fact does not change. Therefore, the controlled grammar should be stored separated from the grammar, in a form which is easily changeable (e.g. as files which can be loaded at runtime).

This means that the Controlled Grammar Verifier uses two grammars, a standard one for the analysis of the input, and a second one (the controlled one) for identification of the controlled language subset. The verification task consists in matching the controlled language with the regular one, and in producing diagnostic information for the non-matching parts.

As a result, a Controlled Grammar Verifier basically consists of four components:

- an **input sentence analyser** which produces linguistic structures of the sentences to be checked; this uses a large coverage standard grammar;

- a collection of linguistic structures which are considered to be **ill-formed** (according to the controlled grammar definition); they are collected in a special repository;

- a **matcher** which matches the input structures with the potential ill-formed structures and flags the deviations;

- an **output** module which produces useful diagnostic information.

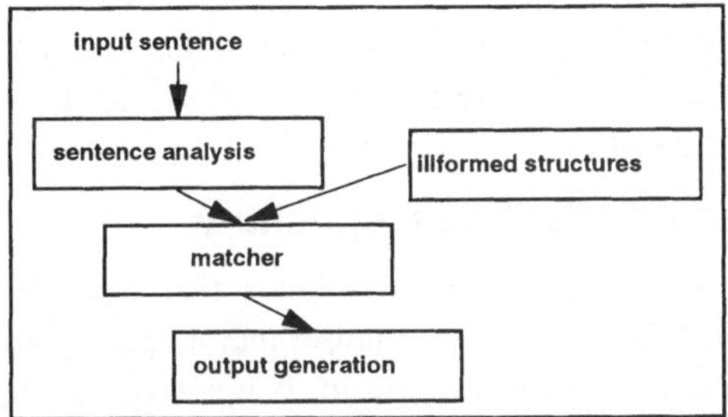

FIG. 2: Architecture of the Controlled Grammar Verifier

3.2   The input sentence analyser

To do diagnostics on linguistic trees, it is presupposed that these trees have been built; i.e. a parser and lexicon must be available which produce these structures.

For a number of reasons, we have chosen the METAL analysis components. The basic reasons are (cf. /THU90b//)

- METAL is available with large coverage grammars and lexica; this allowed us to concentrate on the verification task immediately.

- METAL allows for a flexible grammar approach which, due to its levelling and scoring concept, makes it also usable for grammar checking (concept of fallback rules, cf. /ALO92/).

- METAL is robust: In case of parse failures, partial results can be offered to the diagnosis component. This is due to the fact that METAL uses an (active) chart parser.

However, as a Controlled Grammar Verifier deals with a subset of the grammar, it should be implemented such that the grammar is not touched at all. The only input needed should be a description of the output which the grammar produces;

then any parser and grammar can be used as long as it produces the kind of trees specified.

In TWB, this was the guideline for the implementation: Nothing was changed inside the METAL analysis components to do the diagnosis. The only relevant information was the output of the syntax analysis. This allows the verification to use any grammar and parser which produces the same kind of output structure.

The output structure expected is basically an X-bar syntactic tree, annotated with features. Not all of them are needed for diagnosis. Technically, they are represented as list structures.

3.3   The illformed structures repository

The modular concept outlined above implies that there is no interference between analysis and diagnosis. This implies that no partial analysis results, no special rules etc. are available. As a result, diagnosis can only refer to output structures which means that it must be descriptive: It describes structures which are correct or which are incorrect. This approach makes the diagnostic module very transparent and declarative.

The first task is to reformulate the statements of the controlled grammars in terms of linguistic structures and features: Which structures should be flagged when a sentence must not be "too complex"? Which structures indicate an "ambiguous prepositional phrase attachment"? etc.

The result of this step is a set of structures, annotated with features, the occurrence of which indicates an ill-formed construction.

The next step is to find a representation for these structures. It must be as declarative as possible, which implies two requirements:

- the structures should be stored in files, not in programs, in order to ease testing, but also to change applications (and languages) later on (e.g. from German Siemens Nixdorf Styleguides to English AECMA controlled languages).

- the structures should be declared in some simple language, describing precedence and dominance of nodes, presence and absence of features / feature combinations and values. Any linguist should be able to implement their own sets of controlled grammar phenomena and call them by just specifying their respective diagnostic file.

Both requirements are fulfilled in the final TWB demonstrator; the ill-formed structures are collected in a file which is interpreted at runtime; and the structures are described in a uniform and easy way:

They basically consist of a test and an action pattern. The test pattern specifies a tree (in terms of dominance and precedence, including wildcards) and the presence or absence of some features or values of features. The action pattern consists in adding a new feature-value pair to the top node of the tree inspected; this is used for output generation later on.

This is how the criteria of the controlled grammar could be implemented. It is easy to implement different controlled languages this way as only sets of structural descriptions must be produced and stored in the illformed structure repository.

The prototype of the controlled grammar verification operates on German texts; it subdivides its structures into different areas, following the main linguistic categories:

- sentential issues
- nominal expressions (unclear compounds, appositions)
- verbal expressions (passive voice, complex verb structures)
- lexical issues (wordy expressions etc.)

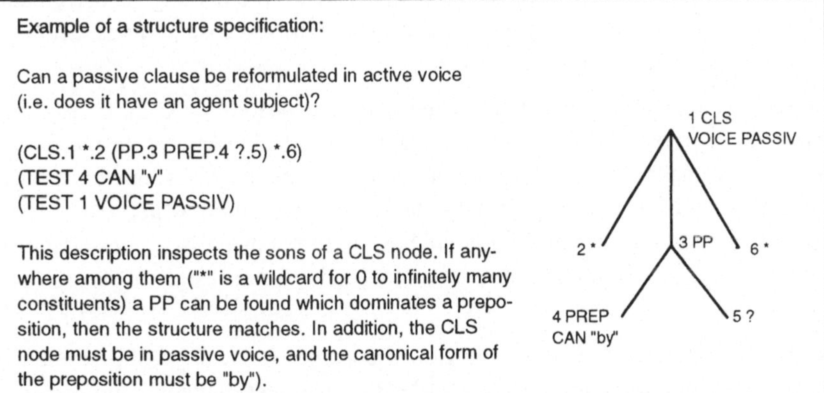

FIG. 3: Example of a tree description

Among the features we have implemented are:

- sentence length

- number of subclauses (not more than three, not deeper than two levels)

- clauses as fillers of verb arguments (Daß er kommt, beweist, daß er lügt.)

- "Satzklammer" (too many constituents between the parts of a verb)

- unclear passive constructions (cases where the use of a "by"-PP creates ambiguities, e.g. in cases of PP attachments)

- unclear PP and conjunction scope, ambiguous references

- unclear compound structures in three-part compounds

- wrong specifier formation in NPs

- word usage (abstracts, wordy expressions)

Some of these issues create deviations of the controlled grammars in their own, others contribute to an overall ratio: There should not be too many abstract nouns, too many passive sentences in the text as a whole. This must be distinguished in the diagnosis module.

In implementing the items just mentioned, we encountered three problems.

The first one was that the level of analysis is different: Some aspects (e.g. the number of constituents between predicate parts) are very much surface-oriented; they may have even disappeared in the parser output (by removing auxiliary nodes in favor of feature structures). Some others, like dealing with thematic roles, require full deep syntactic analysis. This fact sometimes requires additional information in the parser output.

The second one was the fact that some parser decisions must be revised for diagnostic purposes: In case of ambiguous PP attachment, for instance, the parser must finally decide what structure to build. The diagnosis, however, must revise this decision and inspect the potential configurations which could have led to ambiguity (i.e. which rules could have been fed by a given constellation). This means that several possible structures must be checked to give proper information.

Finally, if the parser fails, the diagnosis becomes difficult as the structures expected are not delivered. This could lead to incorrect diagnostic information and influences the reliability of the Controlled Grammar Verifier as a whole.

3.4   The matcher

The matcher is the central component of the verification software. It matches the structures of the ill-formed structure repository with the input sentence, applying the feature and tree structure tests to the input tree. This process has to be done for all subtrees of a given syntax tree recursively (as there may be diagnosis information on all levels of a tree).

For every positive match, the matcher executes the action part of the description, i.e. puts a feature onto the root node of the input tree the value of which indicates the kind of ill-formedness, and gives a hint for the production of the diagnostic information. The output of the matching process is the input analysis tree, modified by some features if ill-formed structures were found.

As a software basis for the matcher, we were able to use a component of the METAL software which performs tree operations.

3.5   The output generator

The last component is the output generator. The diagnostic information must be presented in a way which is easily understandable and optimally usable for the users.

There are two kinds of information to be presented:

As a first type there are sentence specific information flags for ill-formed structures in sentences. This information should be given together with the sentence it occurs in. This could be achieved by either splitting the input text into single sentences and flagging them if necessary, or by writing comments into the document itself (using an additional right column). The latter presentation is preferable but requires some layout information from the document (to find the respective text line) and is therefore restricted to a given editor.

The other kind of information is global and refers to a text as a whole; examples are flags like

- to many passives in a text,
- nominal style,
- overall readability.

This information could be presented in a header of the text as a whole, and could even be represented graphically (e.g. by bar charts).

The existing TWB prototype only supports sentence-based diagnosis and represents it as pairs of <sentence ; diagnosis>. This has to be improved; also, experiments related to a good readability score have to be performed in order to meet users' intuitions on this issue (cf. /KIN81/, /KIN90/).

4   Tests and results

Several experiments have been carried out with the Controlled Grammar Verifier.

In order to test its functionality, a set of test sentences was written which was used for functional tests. They contained phenomena on which the verifier was supposed to react. The verifier was tuned and debugged accordingly.

Moreover, we also analysed several "real life" documents, mainly user manuals in data processing. They consisted of about 220 sentences. The result was that 100 of them were flagged; this turned out to be too much; but 44 just flagged "abstract nouns" (the basis was a guideline that asked authors to avoid abstract nouns). If this was eliminated one out of four sentences was flagged which for the texts chosen was considered to be acceptable. In about 25% of the sentences, a parse failure was reported.

A closer look at the data showed encouraging results: There was a number of cases where the system behaved like expected:

- Läuft das Programm, Daten eingeben
    Sentence contains a non-finite predicate
    conditional clause without subjunction

- Daß er nicht kommt, verhindert er
    Object argument is filled by a clause
    Object is placed before subject
    Double negation because of implicitly negated verb

- die von ihm gekauften Autos
    complex adjectival group

- Reichsversicherungsanstalt
    unclear compound structure

- Reichs-Versicherungs-Anstalt
    --

For three phenomena, evaluation was done, with three problems in mind:

- Does the system flag the correct structures only (i.e. does it overreact)?

- Does the system find all the structures of a type to be flagged (i.e. does it underreact)?

- Does the system's behaviour meet users' intuitions (i.e. can the user understand the system's behaviour)?

First, we checked complex prenominal modifiers. The result was that all cases (13 overall) had been identified and flagged correctly. Examples were

- Dies gilt auch für **die an den Anwender-Computer zu übertragenden Codes** bzw. Zeichenstrings

- Durch diese Strukturierung des Wortschatzes wird die Dialogführung des Benutzers unterstützt, da nur **die für die augenblickliche Phase des Dialoges als aktiv zugelassenen Wörter** vom Gerät angenommen werden

- Er besteht aus **den für die jeweilige Anwendung ausgewählten** Begriffen

In three cases, however, flagging does not seem to be necessary (these structures do not seem to be considered as complex by users):

- chains of two non-modified adjectives:
  vereinfachte direkte Datenerfassung

- participles with non-NP modifiers:
  Die CSE-Spracheingabegeräte sind **selbständig arbeitende** Prozessoren

- conjunctions of simple adjectives:
  Sie beschreiben **akustische und phonetische Eigenschaften** des Signals

This requires tuning of the linguistic descriptions in order to meet users' intuitions.

The prenominal modifiers are a good example to start with as they are easy to describe. The second phenomenon investigated was more complex, namely PP attachment. There were 44 PPs which could create ambiguity problems. After having corrected a wrong structural description of ambiguity constellations, 23 were identified as being ambiguous; among them were the followingy:

- Unter Spracheingabe verstehen wir die **Eingabe von Daten per Sprache** in den Computer.

- Die Operateurkonsole dient hauptsächlich der **Identifizierung des Sprechers beim Laden sowie im Trainingsbetrieb zum Aufbau und zur Aktualisierung** des Wortschatzes.

- Die CSE-Geräte bieten eine sehr hohe Erkennungsgenauigkeit und damit eine größere **Sicherheit bei der Datenerfassung gegenüber den herkömmlichen Erfassungsmethoden.**

- Das Magnetbandkassettengerät dient dem Sichern des Wortschatzes nach dem Training und seinem **Laden von der Magnetbandkassette in den Sprachprozessor im Erkennungsbetrieb.**

Of the rest, 12 were not identified due to a parse failure. This turned out to be a problem for the diagnosis component; we had to improve the robustness of the parser.

In some cases, the system did not meet users' intuitions:

- PPs in prefield position are not considered to be ambiguous as they always form one constituent in German:
  Zur Kommunikation mit dem Anwendungscomputer bietet er als E/A-Interface eine V24-Schnittstelle.

- measure expressions need special treatment: the measure unit attaches to the measure expression and is not ambiguous:
  Die mögliche Eingabegeschwindigkeit kann dabei bis zu **180 Wörter pro Minute** sein.

- directionals nearly never attach to NPs:
  Das Spracheingabegerät bildet aus den gewonnenen Bitmustern einen "Mittelwert", der als Referenzmuster **in den Wortschatz** aufgenommen wird.

This had to be fixed as well.

The third phenomenon tested was wrong NP specifier formation. Here, all 7 occurrences had been correctly identified, e.g.

- Das **CSE 1050-Gerät** besteht aus: ...

The only mistake was the recognition of a complex number:

- einstellbar von **110-19.200** Baud

As a result, the system behaved pretty much as expected; it seems that it is really possible to build a useful controlled grammar verification tool. It needs further tuning, which is not easy in some cases: For instance, how to meet users' intuitions on a "complex sentence"? Is it the number of nodes of a tree, related to the number of terminals? The number of tree levels? The occurrence of certain (e.g. clausal) constituents? This must be tested in more detail.

What turned out to be a problem is the treatment of parse failures (about 25% of the texts could not be parsed). In this case, the diagnosis can be erroneous; e.g. if the system flags an "incomplete" sentence structure, but the incompleteness is only due to the fact that the parser could not find all predicate parts.

The system must not give no information in these cases; otherwise it becomes unreliable from a users' point of view: Users want constant quality of diagnosis, they are not interested in the internal problems of a parser.

Perhaps the system should not try to give diagnosis for certain clause related phenomena, e.g. do not flag a "non finite predicate" in a parse failure sentence like

* Damit koennen C2 Bandmaschinen beschichtet werden:

It still can verify the NP and PP instructions, and others.

This problem requires

* greater robustness of the analysis system's behaviour, like improvement of partial parsing, explicit reference to parse failures, etc.

* some tuning and intelligence of the diagnosis component in cases where parse failures are encountered.

5 System environment

Once the diagnostics component works as expected, the question must be solved in which environment such a component could be embedded.

First, it must be provided that texts are at all importable, which implies problems of editor conversion and deformatting. The controlled grammar verifier uses the text processing interfaces provided by the METAL functions (METAL Document Interchange Format, Plain Text Format); all the converters and deformatting routines are therefore available to the verification tool. This guarantees that text to verify is available at all.

Second, as explained in the first section, the task of verification is more complex than only grammar verification. The complete task refers to several levels and processes, and involves terminology, layout, and text presentation verification as well. The complete task structure looks as follows:

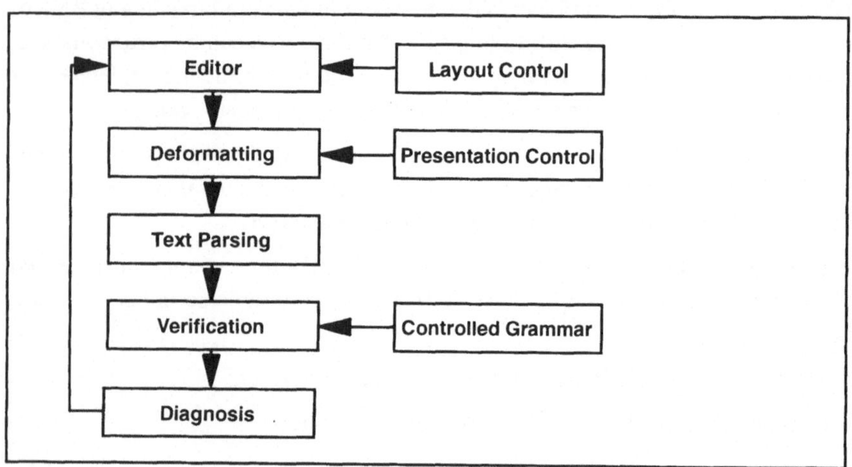

FIG. 4: Task Structure

Third, interaction with an editor has to be determined: If the verifier runs in interactive mode, text portions have to be sent back and forth between editor and verifier. This requires an interface inside the editor which could be supported (those interfaces exist for some editors, like Word or Interleaf, but not for others); it also implies sufficient performance of the verifier, otherwise users must be idle for too long. If the system runs in batch mode with file interfaces, the presentation of the diagnostic information and its links with the original text must be considered.

Fourth, the task must be integrated into other tasks of text control in a documentation environment. To give an example: Users who have checked a text for spelling errors and terminology consistency (which implies dictionary access) should not be confronted with the fact that the verifier does not know words which the speller does know, and vice versa; otherwise, the system as a whole is inconsistent. This implies changes in the overall lexicon organisation, for instance.

Although these aspects do not influence the linguistic core of the controlled grammar verifier, they influence massively its success and acceptance.

6   Next tasks

In order to make the Controlled Grammar Verifier really productive, the following tasks must be performed:

- Port the Controlled Grammar Guidelines to other applications. This should verify if the modular architecture which is based on specification files really works. This task has begun with an external pilot partner.

- Tune the system to find out if what it flags is what human users would flag as well. E.g.: Are all constructions marked as "too complex" by the system also considered to be complex by human readers? This also relates to the number of flags to be allowed in order to be acceptable.

- Improve the quality of the output component: We must be able to refer to the original text in giving diagnostics. E.g. if a sentence contains three large compounds, we must tell the users to which of them the message "unclear compound structure" refers. We also need good text related scores (e.g. for readability).

- Finally, we need a better system integration and a better user interface which allows for the selection of some parameters (do not always check everything) and other more sophisticated operations.

7   References

[1] /ADS90/ Adriaens, G., Schreurs, D.: Controlled English (CE): from COGRAM to ALCOGRAM. Proc. Computers and Writing III, Edinburgh, 1990.
[2] /ALO92/ Alonso, J.A.: The Spanish Grammar Checker. In: M. Kugler, ed.: TWB Final Report. 1992. to appear
[3] /DUC84/ Ducrot, J.M.: TITUS IV System: système de traduction automatique et simultanée en quatre langues. Proc. EURIM 5 Conference, ASLIB, London 1884
[4] /FRE91/ Freijser, J.: "Sentence too long; consider revising." INKtern 1991
[5] /HEL90/ Hellwig, H.: Theoretical basis for efficient Grammar Checkers. TWB Report, 1990

[6] /KIN81/ Kincaid, J.P., Aagard, J.A., O'Hara, J.W., Cottrell, L.K.: Computer readability Editing System. IEEE Transactions on Professional Communication, PC-24/1, 1981.

[7] /KIN90/ Kincaid, J.P., Kniffin, J.D., el al.: The Simplified English Analyzer: A computer aid for authoring text written in controlled English. Proc. of the Topical Meeting on Advances in Human Factors Research on Man/Computer Interactions, Nashville, Tn, 1990.

[8] /KUG92/ Kugler, M., ed.: TWB Final Report. 1992 (to appear)

[9] /MCD89/ McDaniel, B.A.: Principles of Linguistics and Cognitive Psychology Applied to Technical English. Proc. IPCC 1989, Garden City, NY

[10]/SMI89/ Schmitt, H.: Writing Uderstandable technical Text. TWB Report. 1989

[11]/THU90a/ Thurmair, G.: Style Checking in TWB. TWB Report, 1990

[12]/THU90b/ Thurmair, G.: Parsing for Grammar and Style Checking. Proc. COLING Helsinki 1990.

[13]/THU91/ Thurmair, G.: METAL: Computer Integrated Translation. in: McNaught, J., ed.: Proc. of a Workshop on Machine translation, Manchester. 1991

Natural Language Products in Restricted and Unrestricted Domains

Johannes Arz[1], Ralph Flassig and Erwin Stegentritt
Transmodul Software GmbH
Am Staden 18, D-66121 Saarbrücken

1   Natural Language Systems

From a commercial point of view, Natural Language (NL) systems can be divided into two categories:

Natural Language Interface Systems

- Question and Answer Systems

- Programming in Natural Language

- Natural Language Control Systems

Text Processing Systems

- Spelling and Style Checking Systems

- Information Retrieval Systems

- Automatic Translation Systems

- Information Formatting and Presentation Systems

The Natural Language Interface (NLI) category consists of interfaces connecting the user of a system to the system itself. The special capacities of the interface depend upon how sophisticated its treatment of language is. Commercialized products offering Question and Answer Systems include Q&A by Symantec and DOS-MAN by Transmodul.

Text Processing Systems feature the ability to read and write texts. This second category includes Spelling and Style Checking Systems (e.g., Primus by Softex), Information Retrieval Systems (e.g., Spirit of Systex, cf. Chap. 4 and 5, below), Automatic Translation Systems (e.g, Metal by Siemens Nixdorf

---

[1]   also: Fachhochschule Darmstadt, Schöfferstraße 3, D-64295 Darmstadt

Information Systems; Eurotra by the European Community) and Information Presentation Systems (e.g., WIP by DFKI).

NL systems differ from other systems in their capacity to "understand" (analyse or synthesize) language. Yet Natural Language Interfaces must still demonstrate their competitive advantages over other interface products. Text Processing Systems must either compete with existing products or open completely new markets.

Databases can also be accessed by query-by-example. Operating systems function by Graphical User Interface (GUI) or formal language. Text Retrieval Systems can perform indexing using non-linguistic approaches. We see that alternatives to these systems using non-linguistic approaches exist.

One alternative to an automatic translation system is human translation aided by computer support (translator's workbench). Style and grammar checking systems partly replace human work. These systems might very well open new markets.

The questions facing us are: Are these systems as successful as other systems on the software market? And: how should their commercialization best be promoted?

In this context we would like to put forward one initial claim:

Claim 1:

Natural Language Systems have to be commercialized like any other product on the market.

The above claim implies that a certain amount of money must be invested in the promotion of the product in order for it to be a commercial success. (cf. Chap. 3, below). The product must either create a new demand or offer advantages over other products.

We observed that when commercial NL products first appeared on the market, they attracted the consumer because of their unprecedented capacities and seemingly breakthrough features. The consumer, however, has now become more discerning. Today, NL products are no longer sold solely for their novelty, but rather only when the consumer is convinced that his specific user needs can be met substantially more efficiently with NL features. (This is the same turn of events we have observed in the field of expert systems).

An NL product, like any other commercial product, is subject to the usual market laws. Thus we put forward Claim 2:

Claim2:

Promotionary advertising for NL products is - as for any commercial product - essential and legitimate.

Protest is often heard against such marketing strategies. Yet take, for example, television and magazine advertisements for detergents all promising the whitest white ever. Obviously most statements are either only partly true or ambiguous. A slogan such as "My PC understands me" is no more overstated than any other advertisement for a wide variety of quality products.

It is often argued that the advertising of Artificial Intelligence systems must be overly cautious in promising human faculties in their products. They claim that many gullible consumers will prove to be disappointed in the end, and as a result the entire discipline of Artificial Intelligence will be depreciated, researchers and purveyors alike. This fear can be refuted by our claim that advertisers of *any* sophisticated high-tech product (e.g. microwave oven) are faced with this eventuality, yet still continue to be successful in their sales campaigns, never sullying the reputation of the developers of their product. It is a mistake to believe that NL products belong to a special category of product subject to different market laws.

2   Integrated Natural Language Interfaces

Yet even if sophisticated marketing strategies are used, commercial success is not guaranteed. "Even NL cannot make a bad product good, but NL can make a good product better" (Gary Hendrix). The most important requirement of a natural language system today is its ability to be integrated into an underlying core system.
This argument leads to Claim 3:

Claim 3:

Only highly integrated NL Systems can succeed on the market.

This last claim is borne out by the failure of the stand-alone systems, which provide only an NL surface. Today the user demands a combination of NL ability with other forms of interfaces, such as GUI.

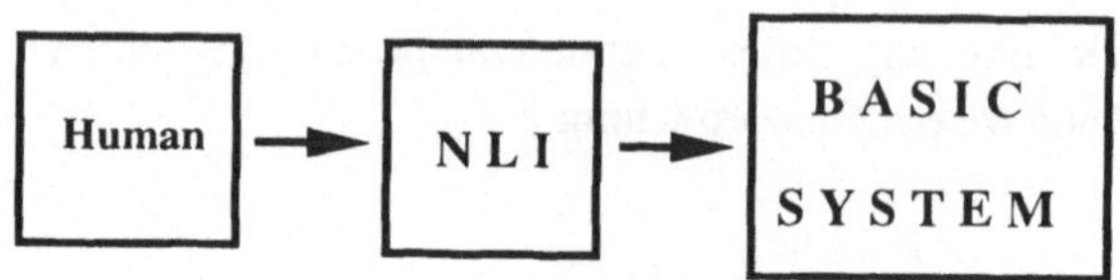

Picture 1: Old fashioned Natural Language Interface

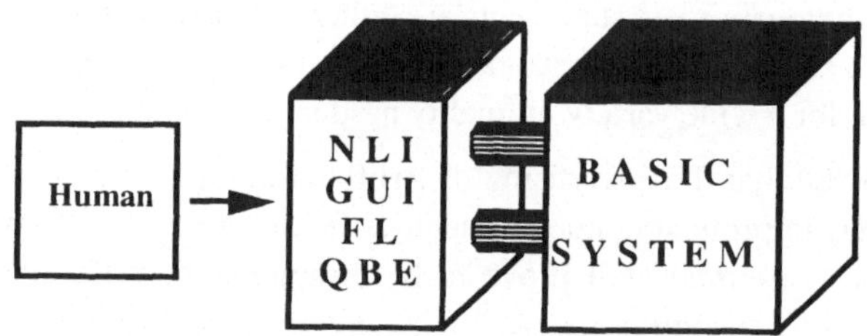

Picture 2: Integrated Natural Language Interface

Examples:

Using the product DOS-MAN by TRANSMODUL in a DOS environment, the user has four ways to view the directory tree diagram:

- NL: "Show me the directory tree."

- GUI: select actions, choose directory tree

- Hot-Keying: a special key stroke (F8 key)

- Formal language (FL): the DOS command tree

The user can choose the type of interaction which is most appropriate for him. He can decide upon one way which he continues to resort to out of habit, or he can progress along different stages, starting with natural language as a beginner, graduating to GUI as an advanced user, and arriving at the use of hot-keys as an expert.

Searching in a Q&A database, the user can choose the following interaction to select a record:

- Query-by-example (QBE): for quick and easy selecting,

- Natural Language: for more complex selecting (aggregate, cross tables, etc.),

- Menu-Based Natural Language (the technique developed in NLMenu by C.W. Thompson): for an inexperienced user.

Q&A and DOS-MAN are distributed through dealer outlets for 1,500 DM and 100 DM, respectively. They are not sold primarily as NL products, but rather as business products with integrated NL facilities.

Although the 1990 initial boom in sales of GUI seemed promising (initiated in the PC domain by the success of Windows 3.0 of MS), users were far from satisfied and their expectations failed to be fulfilled. Lotus 1-2-3, for example, comes with 70 icons ("Icon overkill" as one critic put it). Or try as a beginner to format a floppy disc on MS Windows: you soon become lost in a nightmare of menus and submenus. As a result, many experts have now become more cautious in their optimistic predictions on how such systems can help the user. (To the reader, who disagrees with these statements, we know of a lot of people for who the statements are true. So that proves the need for multiple interaction forms.)

The main problems in NLI are the coverage of the domain and the coverage of the user's view of the domain. These shortcomings of the systems are well known. Knowledge-based systems are limited to a particular scope of capabilities. They are forced to fail at any task even slightly beyond this scope, although the lay user might easily be led to expect a greater concordance between his demands and the systems capabilities.

Of course, formal language interfaces and GUI are also subject to these limitations, yet here the user does not even have the possibility of requesting the system to perform non-available functions. With NLI, in contrast, the user can formulate his request in natural language and has a greater chance of becoming quickly frustrated when the system fails to respond or carries out the command inaccurately.

Faced with the problems mentioned, research institutes and centres will be forced to devise new solutions in order to make natural language interfaces more attractive and responsive to user needs.

3   Success with Natural Language Systems

Success with Natural Language Products as a result of professional marketing can be illustrated by the experiences of two companies working in the NL domain.

Symantec Corporation based in Cupertino, California, is one of the leading software companies in the PC market. One of the most famous products is Q&A.

Take a look at the success story of Symantec. Founded in 1982 with about 10 employees and a backbone of five million dollars venture capital, they invested large sums of money to market their main product, Q&A. At their nadir in 1988, they had fallen to their greatest deficit of about 14 million dollars. By 1991 they had revenues of 116 million dollars, 585 employees at 35 locations and have now become one of the giants in the business in the USA.

Bearing in mind that 1/3 of the revenues are from Q&A, we see that one can indeed achieve substantial earnings with NLI. Yet professional marketing and solid financial support are indispensable.

Milestones of Symantec

- 03/82   Symantec founded by Dr. Gary Hendrix
             (about 5 Mio Venture Capital)

- 09/84   Symantec merged with C&E Software

- 01/87   Breakthrough Software acquired by Symantec

- 08/87   Living Videotext acquired by Symantec

- 10/87   THINK Technologies acquired by Symantec

- 07/89   Symantec's Initial Public Offering

- 08/90   Peter Norton Computing merged into Symantec

- 8/91   Zortech acquired by Symantec

Progress of Symantec

- 12/85 $10,4 Mio. deficit (accumulated)

- 12/86 $2,5 Mio. deficit (accumulated)

- 12/87 $13,5 Mio. deficit (accumulated)

- 03/88 $13,9 Mio. deficit (accumulated)

- 03/89 $10,8 Mio. deficit (accumulated)

- 03/90 $ 9.4 Mio. (Net Income)

Current (3/91)

revenues: 1/3 of revenues from Q&A

- 89/90: $ 74.4 Mio.

- 90/91: $116.3 Mio.

    → (+ 56% to previous year)

TRANSMODUL Software GmbH was founded by Johannes Arz & Erwin Stegentritt in 1984. The main projects in NL include the international Q&A versions (languages: German, French, Dutch, Swedish, Spanish, Portuguese, Italian, Norwegian), DOS-MAN, a natural language interface and help system for MS-DOS, and the German version of Spirit of Systex (cf. Chapt. 4 and 5, below).

Revenues of TRANSMODUL
(70% with natural language products):

- 1989:   DM 400.000

- 1990:   DM 500.000

- 1991:   DM 800.000

This proves that in spite of contentions made by some people one can indeed increase earnings with Natural Language products if astute marketing strategies are used.

4   The SPIRIT OF SYSTEX System

4.1   Characteristics

Q&A and DOS-MAN work only on very restricted domains. SPIRIT OF SYSTEX, a system for free text retrieval, instead is designed to work in potentially any (unrestricted) domain. The following provides an overview of the features of SPIRIT.

a) Free-text Automatic Indexing
    Spirit extracts and standardizes expressions, measuring their relevance against a statistical model.

b) Natural Language Query
    Spirit accepts free format natural language queries. The only knowledge needed is of the document topic. Documents are ranked in order of relevance, according to the key words in common with the query. For large documents, the most relevant pages are displayed first.

c) Dynamic Hypertext Links
    In each query, a statistical model identifies the distance between the query and the database. A semantic match (or near match) is made using the weighting done previously. This provides dynamic hypertext links with other documents or images.

d) Reformulation
    The same idea or concept can be expressed in many different ways. The paraphrasing module therefore transforms the query into all its equivalent forms.

e) Interface to User Programs
    Spirit offers interfaces to management or user programs written in high level languages (PL/1, C,...). These programs can be used both for the generation and modification of databases and for the acquisition of data.

4.2   Processing stages of SPIRIT

The structure of the SPIRIT System and the processing stages are shown in the Picture 3 below.

The linguistic features of SPIRIT include:

*   Morphological analysis and error detection

    With the help of a huge dictionary (full forms) and complementary procedures text-entry in different formats is accepted. French and English versions are today available. A German version will be shipped in 1992.

*   Idiom and compound recognition

    Based on a small idiom dictionary and algorithmic checks idioms and compounds are recognized. Some linguistic ambiguities are resolved by a syntactic analysis with automatic rule modification.

- Standardization

  After removing stop words using either grammatical or morphological criteria, the remaining keywords are standardized.

- Statistical analysis of the database

  Based on the information value of the document in the database, a certain weight is attributed to each word (single term or compound).

Different query types are also possible:

- Natural language query

  The natural language query facility checks for typographical errors and accentuation; searches for compounds and word standardization before SPIRIT looks for relevant documents. The answer is a list of documents in order of relevance.

- Document query

  A document can be used as a query. The answer is a document list ranked in accordance with the words in common with the querying document.

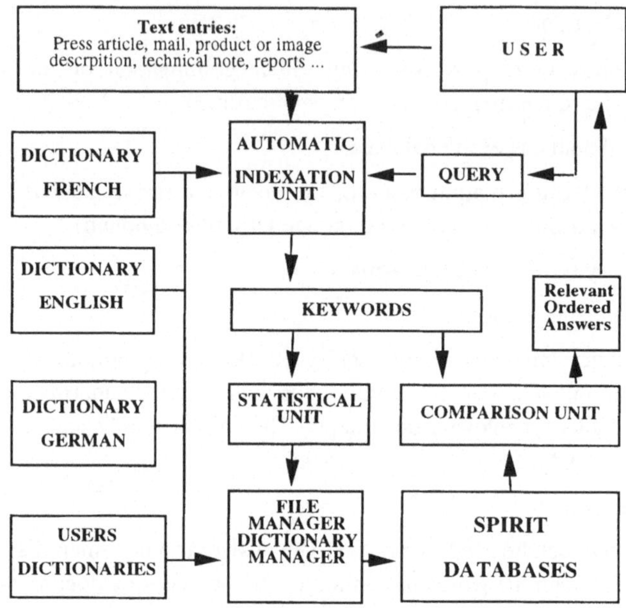

Picture 3: The architecture of the SPIRIT system

4.3   System installations

Some references of SPIRIT (1990) testify to its success:

CEA (French Atomic Energy Authority)

- BRT (Technical Information Office)
  INSTN / Saclay

RENAULT

- Spare-parts management

Electricité de France EDF

- Management of financial and legal departments: National applications of legal cases

TF 1 (French Television Network)

- Automatic indexation of incoming news from news-wires in real time

SNCF (French Railways)

- Accident reports database

DASSAULT (Airplanes)

- Electronic document databases management

GROUPAMA (Insurance Co.)

- Automatic selection of key words connected to DBMS

BULL

- Electronic mail integrated in DBMS
  Management of error messages

5   The E.M.I.R Project

5.1   Goal of the project

The SPIRIT System[2] is used as a basis for multilingual retrieval, developed as part of the ESPRIT initiative of the European Community (Project No. 5312), entitled E.M.I.R (European Multilingual Information Retrieval).

The Goal of EMIR is to develop a "cross"-multilingual (English/French/ German) version of SPIRIT. Queries can be issued in English, French or German, and databases in which the search is performed can be in English, French or German.

---

[2]Most of the linguistic features of the SPIRIT system are now available as the EXTRAKT system by TRANSMODUL.

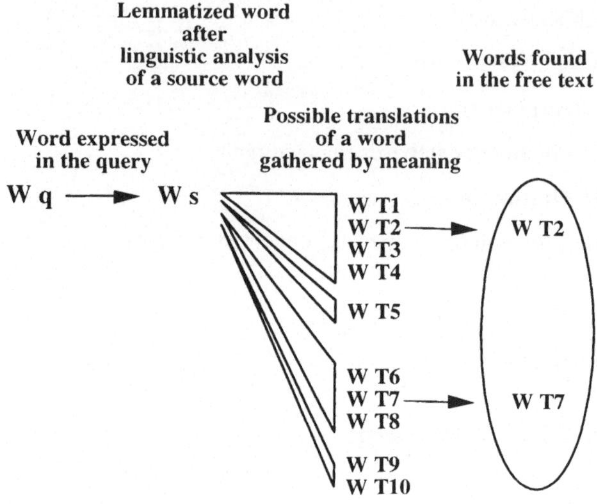

Picture 4: Transfer from Source Language to Target Language

The translation (or "reformulation") of the query is carried out by the system automatically. The reformulation is also performed in the monolingual version of SPIRIT. By this procedure semantic relationships can be treated in an appropriate manner.

The contention - confirmed by initial results of tests with a demonstrator for a bilingual query (French-English and English-French) - is that the "production" of ambiguities in the query will be reduced by the text databases in which the search is carried out. See Diagram 4 above.

The architecture of the E.M.I.R system is shown in the following diagram:

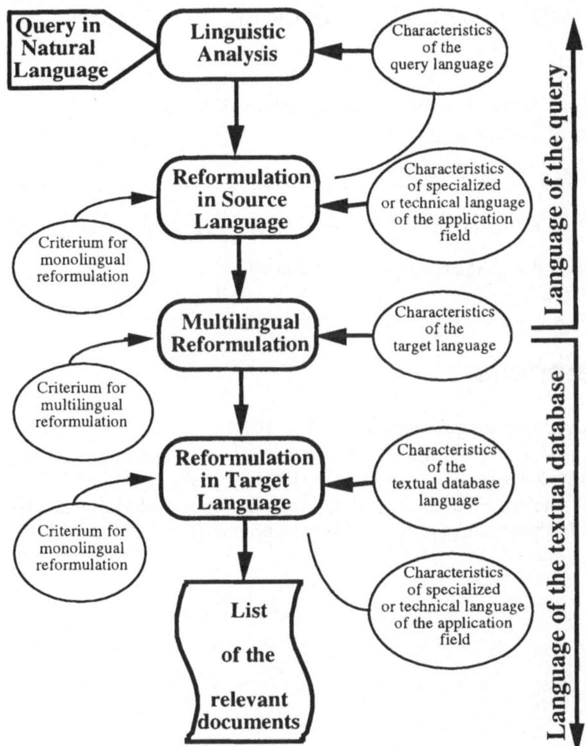

Picture 5: The architecture of the E.M.I.R system

The most important and difficult phase is the transfer mechanism from source to target language. Parts of speech are used to trigger reformulation rules in the same way as is done for monolingual reformulation. Thanks to a monolingual reformulation in the target language, the possible translations replace either the words of the query or the words inferred. Syntactic transformation rules are used to re-build compounds in the target language. A "translation" of the grammatical values (and syntactical structures) of the source language into the grammatical values (and syntactical structures) of the target language is carried out.

5.2   Partners of the E.M.I.R. Project

The partner are:

- INSTN, Saclay, France National Institute of Nuclear
  Sciences and Techniques

- SYSTEX, Gif/Yvette, France (since 1993 Technologies, France)

- UNIVERSITY OF LIEGE, English Department, Liege, Belgium

- TRANSMODUL, Saarbrücken, Germany

6   References

[1] Press releases, Symantec, Cupertino, 1985 - 1991
[2] FLUHR, Christian. "Multilingual access to full-text databases". In International A.I.
     Symposium 90 Nagoya. November 14-16, 1990, Nagoya, Japan.
[3] RADWAN, Khaled, Frédéric Foussier, Christian Fluhr. "Multilingual Access to textual
     Databases". In Proceedings of the RIAO '91 Conference, Barcelona 1991. P. 475-489.

If Language Technology is the Solution, What, Then, is the Problem?[1]

Peter Bosch
IBM Deutschland Informationssysteme GmbH
Institute for Logic and Linguistics
PO Box 103068, D-69020 Heidelberg
bosch@vnet.ibm.com

By a technology I understand a set of technical solutions for a particular set or natural class of problems.

What makes me stop short agreeing that there is such a thing as "language technology", is not so much any disbelief or mistrust in the solutions that are offered under that name, but rather I have some difficulty identifying the class of real problems that these solutions are solutions for.

1   The Human Language Faculty as a Technology

For contrast with what people call "language technology" consider natural language itself. Natural language, i.e. the human language faculty, seems much closer to what I would be willing to call a technology; in this case a "wet" technology, if you wish. And it is a good technology that works reliably. Even though it shares with some man-made technology what we call the banana principle, i.e. it ripens with the customer, this is a process of ripening that reliably yields fair results. Natural language is one of those few technologies that adapt smoothly to whatever purpose they are put. It is user-friendly to the extreme: even the least intelligent specimen of the human race can use it for everyday purposes with no trouble at all.

The set of problems that this technology solves are, in the first instance, problems of *communication.* The transitory meaning-representation in the external physical medium of sound that is used in this technology is, however,

---

[1] This paper has originated from a statement for a panel discussion at the workshop on "Sprachtechnologie und Praxis der maschinellen Sprachvearbeitung" at the Werner-Reimers-Stiftung, Bad Homburg, 27th-29th May, 1991. I wish to express my gratefulness to the Werner-Reimers-Stiftung for their sponsorship and hospitality.

itself capable of being represented persistently and thus naturally allows for other applications of Natural Language, in particular the *storage of information.*

Storage of information and communication are indeed real problems that Nature had to solve at a particular stage of evolution in order to ensure the survival of this race.

2   Problems for Language Technology

But what are the problems that man-made language technology is to solve ? I can see a choice of two basic options:

One is the problem of cloning, in soft- and hardware, a successful and established wetware product of our competitor Nature - up to now the unchallenged leader in this market.

The other problem language technology may be out to solve is to produce periphery, add-ons, and tools for Nature's successful product. And in some cases also fixes are required for Nature's errors or buggy implementations: aids for disabilities in reading, writing, hearing, and speaking - but this is only minor segment in the market.

In either case man-made language technology is dependent upon and secondary to Nature's language technology, and it is Nature who sets the standards. Our products must be Nature Standard or Nature compatible.

3   Nature's Adaptability

In a sense it is easy to make products that conform to Nature Standard, because Nature's technology is ever so adaptable. People who want to use speech recognition for practical purposes today are apparently quite happy to learn to pronounce words separately, with a short pause between them. And for many years anybody who worked with electronic computing gear and natural language didn't know any better than that theses devices knew capital letters only and certainly wouldn't know any letters with funny dots or accent signs over them. Even nowadays there are millions of good people in the non-English speaking parts of the world who see their names misspelled every day for that very reason, and many of them have surrendered to misspelling their own name so as to adapt to a backward technology.

This may create the impression that whatever crap we produce, the customer will get used to it. And I believe this is indeed almost true.

There are only two limitations: One is that the customer will only adapt to a new tool when he has the feeling that it is a useful tool for his problems so that it is worth adapting to it.

The other is a more serious and very natural borderline to the language user's adaptability: he finds it very hard to grasp that language, for the machine, is a purely syntactic thing with no meaning or understanding whatsoever on the part of the machine.

This is not just due to lack of education, but to the very fact that even in science there is no conceptually clear separation between purely syntactic functions of human language and those that require the processing of non-linguistic or world knowledge. The fact of the matter is that Nature's language technology is rather, as hinted above, a language-and-information technology - with language and information inseparably intertwined, while current man-made language technology by and large ignores the information bit and treats language almost exclusively as a syntactic object.

For man-made language technology this means two things:

First, as long as we keep drawing the borderline between language and information in an ad-hoc manner (as we are forced to), practically all our products will yield the notorious 95-percent solutions. This is fine for many purposes, but unpredictable failure, even if only in 5 per cent of the cases, is unbearable in some applications.

Second, all applications are excluded that regularly require knowledge processing, or if you please, AI. And this is an area where I believe that we do have a set of genuine problems that a language-and-information technology still to come could solve: the control and penetration of complex systems by their user, merely by means of the user's mother tongue. But here language is used as a language in the full natural sense: not just a syntactic medium, but a semantically interpreted medium for representing and processing information.

4   Natural Language Interfaces and AI

The area where I see the most promising applications for computational linguistics coming up thus is not an area where we use language now. Rather it is in interaction with machines, particularly complicated ones, and more particularly, devices of information technology, where ergonomic (as well as economic) reasons plead for NL Interfaces, for the processing of spoken as well as written language, for text as well as dialogue.

But there are two problems here:

The first is that although it would be nice to do more with NL interfaces, the market has not yet understood that there is a need for them. In real terms: there is no recognized problem (even though we might have a solution).

The second is that we do not really have a full solution, but only parts of a solution. Computational Linguists cannot build such interfaces. NL interfaces are not just a matter of linguistics, but also of AI: you don't want to give your machine the parse tree or DRS, but you want the poor thing to understand and communicate with the user. But an explanation component of an expert system that was designed with no NL in mind will never talk comfortably to the planner of your NL generation component and parse trees or DRSs are pretty indigestible for current AI systems.

I suppose it is only fair to admit that there is no technology available in this area. Isolated solutions and many brilliant ideas in research: yes. But this is not what I mean by a "technology". Even conceptually we are far from a good understanding of the relation between language and knowledge and even further from an implementation of such understanding in terms of linguistically sensible Knowledge Representation Formalisms.

This is, however, the direction which I believe we ought to go. Not only because it is an exciting area to work in, but also because here we are working towards the solution of real problems for which there is a market.

5   Technology Creates Practice Creates Technology

What I am saying, I suppose, is rather trivial: new technology is created in tandem with a new practice. As long as there is no such practice and there are no products that this practice requires, all our linguistic engineering and computational linguistics is a bunch of nice and fancy ideas, but has little to do with technology.

One could go one step further and not only dispute the existence of language technology but also dispute that it has any right of existence. For the simple reason that there is neither a natural class nor even a significant set of practical problems that it could solve.

The class of problems I mentioned, one might argue, is catered for by what we call information technology and it is in this context where linguistic solutions are required, but as an integrated part of information management, as part of Knowledge Based Systems, and not as free floating parsers.

6   Market-Driven Research

Sure enough, we will never develop a technology in this area worthy of the name, unless we make the market believe that we already have a technology. In a market-driven economy (with market-driven research) we are practically forced to claim that we can turn, if you will pardon me, shit into gold in order to be given a fair chance to prove that we can turn zinc and copper into brass.

In this sense then, I am, after all, glad to pronounce that I believe in the existence of language technology. I only ask your permission to modify this credo slightly: I believe in language-and-information technology. Here we do have a genuine set of problems to which this technology will - in due course - provide the required solutions.

7   But This is all AI and not Really Linguistics

True. And I have also ignored, except for a very short mention, what I called above the area of periphery, add-ons, tools, and fixes to Nature's language technology. But this is exactly the area of which I am claiming that there is neither a natural class of problems nor do the solutions that are available to the various odd and isolated problems in this area have very much in common that would justify calling this motley bunch of tricks a "technology".

This would all be different if we had at our command something like a generic natural language engine that incorporates the best insights of linguistic theory and can provide solutions for linguistic problems on this basis. The fact that no such natural language engine is available is due, next to the enormous difficulty of the subject matter, to the problem of separating, within the functioning of human language, the genuinely linguistic contributions from the contributions of the conceptual machinery and world knowledge, i.e. the problem of separating language and information. I am not sure that this problem can be solved. But what we can do is try to come to a better understanding of the problem, i.e. a better understanding of the interactions of language and information in the ordinary functioning of human language. - I don't care much whether this study is conceived of as part of Cognitive Psychology, Linguistics, or AI. But it is the problem we must focus on if we ever want to get anywhere near something worth being called Language Technology. It is only on the basis of this understanding that we can develop the concepts that can eventually provide generic solutions - and hence a technology in the proper sense - rather than a motley bunch of tricks. The step that is required is comparable to the step from Alchemy to Chemistry.

"Natural" Natural Language-Based Human-Computer Interaction

Oliviero Stock
IRST- Istituto per la Ricerca Scientifica e Tecnologica
38050 Povo,Trento, Italy
stock@irst.it

1   Toward an ecology of human communication with machines

What does it mean for a human-machine interaction paradigm to be natural? Certainly it means to take into account what man has developed filogenetically as his devices for interacting with his similars: natural language in the first place. We are well aware that natural language is not only the main vehicle for such communication, but also that it has the dual role of being the device through which we organize our thought. This makes natural language communication a window on the mind as many human scientists have told us so convincingly, but also implies that natural language communication does not need "cognitive transducers". If often we do need some conscious phase in organizing what we want to communicate, this depends on planning our communicative actions, deciding what we want to say and optimizing our rhetoric strategies; it does not depend much on the lower level of expressing ourselves through words. So, for human-machine interfaces, natural language has an unmatchable potential. On the other side we are facing an extraordinary revolution that touches all levels of our life. Computers and telecommunication are really part of our ecological system.

From the sociological point of view, in the history of mankind never was there such a rapid spread of technology. But even more important, it is a technology that is perceived fundamentally as a two-way interface to an abstract blackbox. Therefore it is bound to be seen as an extension of our human capacity, both in the sense that it makes distances disappear and information be made available, and in the sense that we become used and adapted to the technological features of the interface. What is characteristic of computers, and not of human-human biological interfaces, is that: a) they have a potentially large bandwidth of communication with humans, in particular if we consider their dynamic graphical capabilities combined with other means; b) they provide a visible context, represented on the screen. The transmitted things in the two directions are there, on the same medium, and unlike written paper-based communication the physical medium can be made active. Things can be selected, changed, used for new input. With the concept of direct manipulation even the distinction between the notions of input medium and output medium is blurred.

A new way of interacting, a way that I would still call natural or ecological, even if not just biologically based, is slowly seeing the light. This way of interacting will be centered on natural language but in a new creative way that exploits the possibilities of the computer. The problems involved are tremendous, if we want computers to be able to communicate with us naturally. The questions now are: Where are we? Do we have a natural language technology, such that we can move along this path? Do we have innovative even if initial ideas and prototypes to begin experimenting these ideas?

2   Natural language technology?

You can talk about technology along some different dimensions:

a) when there is a well defined result that a basic technique enables;

b) when for a technique the formal properties have been understood out of which one can foresee the advantages and the limits of the technique (even if not all verified in experimental systems);

c) when the scientific community refers to the technique just by naming it and clearly distinguishing it from other ones;

d) when the technique is considered a building block, efficient enough and available to be used without modifying its internal characteristics.

I think that the term "technology" is appropriate for some areas of Computational Linguistics (CL) and absolutely not appropriate for others. Many aspects of CL have the character of craftmanship, and rightly so, because they permit the development of ideas in a loosely constrained way. Other areas, especially the ones studied for a longer time, enjoy many of the "technology" aspects. Still other areas are simply not like this only because of the inefficiency and lack of organization and seriousness of the Computational Linguistics community.

I maintain that a typical craftmanship theme is the computational pragmatics subarea, and perhaps the craftmanship approach should be extended in this phase so that novel ideas emerge and new experiments are conducted.

For the parsing area we can talk of technology: the community has evolved some solid techniques such as chart parsing [Kay 1980] that are really a firm point for whoever wants to build a natural language system. This was not the situation ten years ago. The convergence of the community on a formalism, with a clear orientation toward declarative representations, has also contributed to the technological development of this area. One very relevant consequence is the consolidation of bidirectional representations, that can be used both for parsing and generation.

Yet, for a field like computational linguistics that has to deal with large amounts of data, it is appalling how little organization there has been so far and how little has been done by those supposed to work for producing sharable dictionaries, usable corpora, multilingual materials etc. The "reusable resources" buzzword has been little more than a joke, bitterly emphasizing an objective problem for the application oriented field. An effort in evaluating natural language systems is similarly very important and really just beginning to take place.

I do not intend to make a state of the art report here, so I shall just very briefly say a couple of words about other critical areas. Semantics is steadily improving, and certainly there is a very long path on which it will proceed. Knowledge Representation has not progressed as we would have needed, also because it went away from the natural language community that was its parent and took care of its infancy. In recent years almost all efforts have gone into proving formal properties of the various approaches, a very important aspect, but one that should not be disjoined by the practical needs and uses of techniques and sytems [Doyle & Patil, 1991]. New promising areas are emerging, among which the integration of symbolic and statistic methods, but the area of generation is perhaps the one in which more progress has been made recently. In fact ten years ago it simply almost did not exist at all! This does not mean there exists a solid technology. It is an area strongly influenced by a couple of schools, but where creativity has some larger space than in language understanding.

Perhaps it is also important to complement these notes on computational linguistics with some notes on computational linguists. I believe a lot of us work without understanding what we are doing, and, equally negatively, without doing enough to make outside people understand the potential of this area. That menus have made applied natural language processing research useless is a stupid statement that we hear very often (e.g. AI Trends, 1991). If this can happen even within the Artificial Intelligence community, then there is something wrong on our part. It is our duty to become a group influential in the outside world, exactly because our experience is such that we can have a role in casting important aspects of the society of the future and because we will not proceed much in isolation.

3   Steps toward an innovative modality

The question now is: Are there innovative ideas and possibly advanced prototypes to experiment with these ideas? I claim that a first important thing to acknowledge is that generally speaking applied natural language processing research should do away with the teletype approach. There is no need, apart from very particular applications, to see the channel of communication between the user and the system as a narrow and constraining device. The advantage of integrating multiple media in output is obvious, for instance, to explain sequences of operations or to display the status of a complex process. Recently some projects of multimodal information presentation have started to combine dynamically graphics and language in output [Feiner & McKeown 1990, Wahlster et al., 1989].

Similarly for input, pointing to images on a screen may individuate the objects involved in some desired action [Arens, 1989, Wahlster 1988, Hollan et al., 1988, Cohen et al. 1989]. Nowadays quite a popular technology is hypertext, a simple idea that we are beginning to use in many practical systems. The generalization of the idea of hypertext to multimediality gives rise to the concept of *hypermedia*: "a style of building systems for information representation and management around a network of multimedia nodes connected together by typed links" [Halasz, 1988]. Hypermediality has opened interesting perspectives on the problem of accessing loosely structured information. Hypermedial systems promote a navigational, explorative access to multimodal information: the user, browsing around the network, is at the same time both exploring the network and searching for useful information.

Hypermediality is a good partner for offering the possibility of realizing intelligent interfaces that amplify capabilities we have in nature (as opposed to trying to reproduce them exactly). I believe that the integration of natural language processing and hypermediality is a very promising approach for at least two reasons:

1) The combination of the relative freedom provided by a natural language interface (with the power of making complex and precise requests and answers) and a visual presentation (with direct manipulation of all kinds of objects) of some organized subdomains has immense potential impact. Beside this, the user can interleave precise requests with concrete exploration of "the surroundings" of her focus of interest. It should be noted that among many interesting aspects, this approach overcomes the typical problems of disorientation and of the cognitive overhead of having too many links in an hypermedia network.

2) The approach is feasible even without a dramatic and unrealistic progress in our understanding and modelling of natural language processes. We must take into account these facts:

   a) the user may have at her disposition the possibility of referring explicitly (by a pointing gesture) to all objects as long as they are on the screen and use them in her requests;

   b) the user does not rely uniquely on the complete understanding of her goals by the system: she can also find out at least some information through direct manipulation;

   c) the user has an explicit model of what information is available;

   d) up to a point, images can be given in output even if they do not have an accurate internal semantics for the system.

All of these aspects result in a simplification of the natural language pragmatics side, the most difficult one to deal with in principle. Limited user modelling emerging in the course of the information seeking natural language dialog may provide a minimal but sufficient means to guarantee a substantial global environment habitability, at least in typically "explorative" and "individually creative" domains.

Of course a sine qua non condition underlying all this remains the development of a fairly sophisticated capability in several components of a natural language processing system - one that is realistic given the current state of the art.

In general, I believe the situation will become more and more common in which a user may want to interact through a dialog with the system referring explicitly to the deictic and linguistic context, incrementally focus in order to obtain punctual personalized information and at the same time the possibility of accessing standard, browsable information through the same generated text.

At IRST we have experimented with these ideas in building a large prototypical dialog system that integrates natural language and hypermedia: AlFresco [Stock 1991][1]. AlFresco is an interactive system for a user interested in frescoes. It runs on a SUN 4 connected to a videodisc unit with a touchscreen where images about Fourteenth Century Italian frescoes and monuments are displayed. The system, beside understanding and using natural language, shows images and combines film sequences. Images are active in that the user may refer to items by combining pointing on the touchscreen with the use of linguistic demonstratives; for example the user can point to a detail of a fresco and say "can I see another painting representing this (+pointing gesture) saint ?" The dialog may cause zooming into details or changing the focus of attention into other frescoes. Also, the systems linguistic output includes buttons that allow the user to access images and an underlying hypertext. The overall aim is not only to provide information, but also to promote other masterpieces that may attract the user.

The knowledge of the system is represented through:

- a knowledge representation language used for defining everything the system can reason about: frescoes, monuments, painters, contents of frescoes, towns etc. and providing the base for AlFrescos deductive inference capabilities;

---

[1] The following people have contributed to the development of AlFresco: G. Carenini, F. Cecconi, E. Franconi, A. Lavelli, B. Magnini, F. Pianesi, M. Ponzi, V. Samek Lodovici, C. Strapparava

- a NoteCards [Trigg et al. 1987] hypermedia network containing unformalized knowledge such as art critics opinions on the paintings and their authors.

The system is based on several linguistic modules such as:

a) a chart-based parser able to deal with flexible expressions, and in particular idiomatic forms and some kinds of ill-formed input;

b) a semantic analyzer able to disambiguate the sentence in the given domain through interaction with the parser;

c) a component that builds the logical form interacting with the Knowledge Base;

d) a topic component that takes into account also deixis (references to images);

e) a pragmatic component, substantially based on a model of the interest of the user;

f) a natural language generator that takes into account the users interest model.

The generated output is in the form of a hypertextual card: the text is enriched with dynamically generated buttons that the user can click to get more information and explore "the surroundings". In fact, the generated texts are immersed in a pre-existing hypermedia network, that therefore allows to the user to browse around, integrate with other information the information provided by the system on the basis of what she seemed to want and possibly find some other stimuli for further requests.

We have experimented with the same paradigm for MAIA, an experimental platform of the integrated AI project under development at IRST. MAIA includes a concierge-workstation that interacts with a visitor to IRST through dialog and provides information about IRST activity, and a porter-robot that moves in the institute. The structure of the information the user may seek is different and so are the media involved, especially as far as images are concerned. Yet we are going to exploit the potential of interacting with real world images taken from the robot while moving around....

4   Conclusions

It is hard to make predictions, especially about the future (Niels Bohr), but I think that the integration of hypermediality with natural language processing technology opens a wide range of new perspectives for the development and deployment of intelligent interfaces. I believe that a potentially very powerful concept is emerging, one that goes beyond reproducing what we had "in nature" in human-human biological communication. An ecological view of communication between humans and machines also requires serious behavioural and experimental studies: the time is ripe for a contribution from applied human sciences.

5   References

[1]   AI Trends, Editorial, February 1991
[2]   Arens, Y., Feiner, S., Hollan, J., Neches, R. "Proc. of the IJCAI Workshop on a New Generation of Intelligent Interfaces", Detroit 1989
[3]   Cohen, P.R. et al., "Synergetic Use of Direct Manipulation and Natural Language", Proc. CHI 89, Austin, Texas, May 1989
[4]   Doyle J., Patil R.S., "Two theses of knowledge representation: language restrictions, taxonomic classification, and the utility of representation services", Artificial Intelligence, Vol. 48, No. 3, April 1991
[5]   Feiner, S. and McKeown, K. "Coordinating Text and Graphics in Explanation Generation", Proc. AAAI-90, Boston 1990
[6]   Halasz, F.G., "Reflections on NoteCards, Seven Issues for the Next Generation of Hypermedia Systems", Communications of the ACM, Vol. 31, No. 7, July 1988
[7]   Hollan, J., Rich, E., Hill, W., Wroblenski, D., Wilker, W., Wittenburg, K., Grudin, J., "An Introduction to Hits: Human Interface Tool Suite". MCC, Tech. Rep. ACA-HI-406-88, Austin, Texas USA, December 1988.
[8]   Kay, M. "Algorithm Schemata and Data Structures in Syntactic Processing", Technical Report CSL-80, Xerox Palo Alto Research Centers. Palo Alto, California, 1980.
[9]   Stock, O. "Natural Language and Exploration of an Information Space: the AlFresco Interactive System" Proceedings of IJCAI-91, the Twelfth International Joint Conference on Artificial Intelligence, Sydney, 1991
[10] Trigg, R.H., Moran, T.P., Halasz, F.G., "Tailorability in NoteCards". In Proc. of Interact 87 2nd IFIP Conference on Human-Computer Interaction, Stuttgart, 1987
[11] Wahlster, W., "User and Discourse Models for Multimodal Communication". In J.W. Sullivan and S.W. Tyler, eds., Architectures for Intelligent Interfaces: Elements and Prototypes, Addison-Wesley, 1988
[12] Wahlster, W., Andrè, E., Hecking, M., Rist, T., "WIP: Knowledge-based Presentation of Information", Report WIP-1, DFKI, Saarbruecken, 1989

The Role of Evaluation in Language Technology

Tom C. Gerhardt
CRETA Luxembourg

1 State of the Art in Language Technology

1.1 Language "Technology"?

Spelling checkers, natural language interfaces and machine translation systems (MT) have something in common: they belong to the group of natural language processing systems and there is no doubt, that the theoretical basis of these systems is called computational linguistics. This basis has its roots in linguistics, i.e. in formal linguistics, in mathematics and computer science. Furthermore, there are influences from sociology, psychology and the interdisciplinary fields among these five. Computer science with its subfields hardware and software is the only candidate which could contribute to a technology.

A technology implies the description of procedures and methods in order to form a product. Hardware is a classical product type where the capacity of all of its parts as well as of the product as a whole is well described. The capacity can be tested, compared with the predefined characteristics and valuated. Software is a different type of product as it is not composed of physical parts. Software consists of programs, which are sets of instructions for calculation. Together with appropriate data these programs should produce certain results. In the case of hardware it is such that the capacity of parts can be tested with regard to their function and it is possible to measure function and size in electrical or mechanical measures. In the case of software the measures for hardware do not fit. It is felt that in the same way as the term product is used in an abstract way the measures have to be abstracted. It is obvious that a more abstract type of product is obtained through different production methods and procedures compared to hardware products.

Language technology (LT) is a conventionalized term and we think that the technology is derived from the software technology enhanced with linguistic know-how, i.e. computational linguistic know-how. As a consequence we think that language technology is the application side of computational linguistics.

1.2 Strengths and shortcomings of today's language technology

Research and development in the field of machine translation and machine aided translation (MAT) started after the World War II. Computer science and formal linguistics did not exist. It was applications which led to the foundation of the sciences (and university institutes) forming the basis of today's LT (e.g. an institute for computational linguistics in Saarbrücken was founded due to the fact that the German part of the European MT project Eurotra was settled in Saarbrücken). M(A)T systems are complex applications the development of which took five to ten years in the past. During the development of a system it is nearly impossible to introduce results of recent research without major changes to the system architecture. As a consequence most systems are outdated from a scientific (and technical) point of view when they are ready to be marketed or when a project comes to its end respectively.

Due to restricted machine power and storage capacity it was necessary to design very compact software packages in the past. Linguistic rules and data were integrated into the software instructions treating the linguistics. Modularity, separation of programs and data, enlargability and other kinds of quality criteria of today's software products are still on the list of demands of marketed M(A)T systems.

Today's methods, formalisms, tools and the logical, analytic way of thinking of computer science are introduced increasingly in computational linguistics with direct consequences for language technology and we think that these are the strengths of today's LT.

The shortcomings are felt to be twofold. Lots of formalisms and procedures respectively were designed for natural language processing, but evaluations proving their fields of application and limitations do not exist (due to linguistic problems?) and it seems to be nearly impossible to describe linguistic phenomena in an adequate way (formally), because only those phenomena, which are formally described, can be used in CL-applications.

1.3   Rate of progress

Concentrating on the public sector on a European level it is amazing to observe the rapid growth of funding in the sector of machine translation and `language industry of Europe'. The EC's translation service is undertaking own projects to improve translation speed and quality. Several EC programmes are incorporating action lines to meet linguistic problems, i.e. multilinguality.

Funding is correlated with public awareness, which has grown reasonably during the last ten years, but even though, funding is totally unacceptable compared to the complexity of computational linguistics.

One stumble stone for progress is the turn-around time of projects. Compared to R&D projects in the seventies project periods were shortened and the workforce per project reduced. This puts more pressure on the projects and leads to small scale experiments which disallow for generalization.

The progress made by the industries could have been better, if the hardware situation had been stable, but a big part of the innovative power of the industries was bound by porting software from one hardware platform generation to the next, from 8 bit to 16 and to 32. It is hoped that this situation can be overcome with standardized software platforms and user interfaces on which the software can be based on.

2   Design and Evaluation

2.1   On the need of a design/evaluation methodology

Within this context we think it is inadequate to discuss the question of the existence of a science or study of methods (German: Methodologie) or the question of the existence of a methodology (German: Methodik) of computational linguistics. So should the question be reduced to design/evaluation methods?

A method is always based on a theory or on parts of different theories, that is on objective criteria. Several methods should, if they are not competing against each other, be classified with regard to the methodology in order to reach their aim - the modification and/or realization of the reality. But the method itself does not indicate the approach to be taken with regard to the data but characterizes a principled procedure.

These arguments show that a discussion on the theory of methods does not contribute a lot to solutions concerning design and evaluation. We want to make some observations concerning design and evaluation, try to find parallels between both and discuss problems and properties of evaluation.

2.2   Some observations concerning design and evaluation

Design and evaluation are two steps in the genesis of a software product. Design is the step that follows after the development of an idea. The aim of design is to realize the idea of the final product. Its function can be described as a process preparing the development of the software. The instruments of this process are criteria, approaches, methods and procedures.

Evaluation is a step that ideally follows all steps of product planning and development. The aim of evaluation is to verify the goals envisaged in the process under consideration. Its function can be described with the word control. The instruments of this process are criteria, approaches, methods and procedures. Unfortunately most evaluations take place at a very late state in the product planning and development procedure with the result that a multi-purpose evaluation is too complex to answer the criteria in an appropriate way.

2.3   Parallels between design and evaluation

The development realizes the design and the evaluation checks the appropriateness of the development. The evaluation verifies the correctness of the assumptions of the design process but only if the design was based on explicitly stated formal criteria.

If this definition of the relation between design and evaluation is correct, it is obvious, that they share a set of criteria. Concerning methods and procedures it is much more difficult to judge on similarities, though it can be assumed, that the underlying principles ought to be similar. This assumption is based on the results of an internal Eurotra paper produced by a special interest group on evaluation (Copeland 1991) describing the procedure of evaluation with its conditions.

The evaluation process is subdivided into four steps:

-   definition of goals and instruments
-   data collection
-   testing
-   validation

The first step, the definition of goals and instruments delivers the framework for the evaluation. It describes the subject of the investigation, the criteria under which the tests have to be performed as well as the criteria against which the subject has to be tested and the test data. It defines the procedures and methods and discusses the motivations for this approach. Data collection means the procedure of selecting the data to be tested under the restrictions given in the description of the test data. If the test is a linguistic test of the syntactic analysis of a machine translation system, the test data subdivides into input data and output data. The input data, in the current literature called 'test suites' (Flickinger 1987, King 1990, Rinsche 1993), are composed of sentences or parts of sentences selected under linguistic aspects in order to test the treatment of linguistic phenomena in various occurrences. The output data are formed by the results of the subject of the evaluation process, e.g. formal descriptions of the analysed test suites.

The test procedure is defined in the first step. Testing means to perform the tests and to check the completeness, formal correctness of input and output data (no spelling errors etc.) and other test stabilizing criteria that shall guarantee the possibility to repeat the test with identical results.

Validation is the keyword for comparison. We mean the comparison of data with data, data with criteria and the results of comparisons with measures. One of the most important conditions of validation is the comparability principle; it is not allowed to compare units which are not of the same kind, and the absence of a unit must not be compared to the existence of a unit. This seems to be trivial but a closer look to an evaluation of complex systems will show the necessity of a strict handling of such rules.

After these four steps the evaluation procedure is finished though the users of the results, e.g. persons financing or engaging the evaluation will interpret the results and they may take into account those items which dropped out from the validation because of the comparability principle.

Evaluation and design do share a good portion of similarities with regard to the underlying criteria, but the implementation of criteria needs different procedures than the evaluation of those.

2.4   Problems and properties of evaluation

If there were a methodology for evaluation it would describe the scope of evaluation (Mann 1981) and hint to the problem, that an evaluation supposed to cover a complex system composed of several subsystems might not cover all subsystems well. A methodology would take into account different types of consumers, financiers, (end-)users, interested parties or whatever the recipient of the evaluation is called.

A quite conservative approach is to evaluate natural language processing systems as ordinary software. But even the evaluation of spelling checkers - an unproblematic tool in the prevailing opinion - is problematic. Concerning the speed of such a tool it is hard to judge, if thirty words per second is fast or slow, especially if other systems with a similar performance do not exist. Additional problems occur, if e.g. one of several systems is more accurate but slower and more expensive. If a cost/benefit calculation is made, what are the measures to compare with? Is it adequate to compare the salary of a well educated secretary with the price of a spelling checker plus the salary of a less well educated secretary? In the case of the comparison of machine translation with human translation is it adequate to add the prices of raw machine translation and of post-edition and compare it with the price for the translation of a human translator?

These questions are pointing to another problem: there are cases (we found many more than these) the answers to which must not be generalized. In other words, the results of an evaluation may only be used for the intended purpose of the evaluation.

To mention the positive aspects design and evaluation of MT systems is finding growing interest not only among scientists. In 1989 a group of MT users was constituted in Germany in order to exchange know-how and experience in the field of MT and MAT. The group grew to a European MT users group with regular plenary sessions in order to define user viewpoints with regard to natural language processing tools. As the members of the group are based in enterprises, administrations and universities representing the groups system user, designer or sponsor it is hoped to arrive at better products and projects. The members of the MT users group are trying to support the field while working in smaller groups under the headings 'quality criteria', 'vocational training', 'introduction procedures' and 'secondary tools'. One of the results of the working groups is a list of evaluation criteria concerning ergonomics, computer technology, management and linguistics. Case studies will prove the usefulness of such a list and they will spell out the conditions under which the criteria may be applied.

A problem of evaluation is the fact that most evaluations are black box evaluations where input and output is compared with each other and with the best possible human translation. These empiric approaches based on direct experimentation with samples of translation units normally are not accurate enough, because the samples are not chosen under representivity criteria, the samples are too small and the measures for comparison are not well defined. In addition most linguistic tests are very much grammar centered. The results highlight just one of several aspects the client of an evaluation is interested in.

At the end of this very brief contribution some more principles shall be mentioned:

- The more complex an evaluation is, the more complex is its interpretation.

- If you can avoid an evaluation, don't evaluate!

- If you don't know the user of an evaluation, don't evaluate!

- System evaluations benefit principally system insiders.

- Evaluations try to find good reasons to invest more money.

A quotation from 'Professional Foul', a play by Tom Stoppard:

> *"The likelyhood is that language develops in an ad hoc way, so there is no reason to expect its development to be logical".*

3 Future developments in language technology

Research and development in the field of machine translation showed that growing complexity of applications also means increase of wrong hypothesis. Negative results of such research and development activities tend to negatively influence the whole scene; especially the sponsoring administrations are very sensitive in this respect. This means that attacking highly complex applications is felt dangerous.

In many cases the results of research and development in highly complex areas are pretty poor compared to the price of the projects. But most of these projects are also addressing political aims, which are satisfied better, than the scientific ones. Concerning operable output the laboratory prototypes tend to be ineffective and unproductive.

Research and development projects in highly complex areas seem to stimulate the scene and multiply the effort invested. Like a steam engine, which is not useful in itself, but only if it pulls a train, complex application areas may prepare the ground for political decisions supporting this kind of projects. A number of spin-offs can be expected having a major impact on industries (far more than the laboratory prototypes available at the end of highly ambitious research). These are the reasons to consider very complex applications in research and development desirable.

4   Conclusions

- LT is felt to be an integral part of computational linguistics; a genuine technology does not exist yet.

- LT made sufficient progress, but funding and linguistics of CL are lagging behind.

- The information exchange between administrations, enterprises and research institutes started recently and needs further improvement.

- Design/evaluation criteria and methods are underdeveloped.

- Future developments in LT need to attack highly complex application areas to stimulate the scene; useful spin-offs and investments of the industries will be the benefits as well as joint efforts to overwhelm language barriers.

5   References

[1] Anwender-Arbeitskreis Maschinelle Übersetzung (1991): "Evaluationsliste (list of evaluation criteria)". Internal report produced by working group 3: Qualitätsvergleich. Bonn, Luxembourg.

[2] Bátori, Istvan S.; Lenders, Winfried; Putschke, Wolfgang (Eds.) (1989). "Computational Linguistics. An International Handbook on Computer Oriented Language Research and Applications". Walter de Gruyter. Berlin, New York.

[3] Copeland, Charles; Gerhardt, Tom C.; Havenith, Roger; Joergensen, Lisc; Ripplinger, Baerbel; Roniotes, Celia (1991). "Recommendation on Testing and Evaluation". Internal Eurotra report. Luxembourg.

[4] Flickinger, Dan; Nerbonne, John; Sag, Ivan; Wasow, Tom (1987). "Toward Evaluation of NLP Systems". Forum of the ACL - Evaluating Natural Language Processing Systems. Palo Alto.

[5] Guida, Giovanni; Mauri, Giancarlo (1986). "Evaluation of Natural Language Processing Systems: Issues and Approaches". Proceedings of the IEEE, Vol. 74, No. 7. 1026-1035.

[6] King, Margaret; Falkedal, Kirsten (1990). "Using Test Suites in Evaluation of Machine Translation Systems". Proceedings of Coling 1990, Vol. 2, 211-216. Budapest.

[7] Mann, William C. (1981). "Selective Planning of Interface Evaluations". Proceedings of the 19th Annual Meeting of the Association for Computational Linguistics, Stanford.

[8] Rinsche, Adriane (1993). "Evaluationsverfahren fuer Maschinelle Uebersetzungssysteme - Zur Methodik und experimentellen Praxis". Commission of the European Communities. Report EUR 14766 DE. Luxembourg.

[9] Sondheimer, Norman K. (1981). "Evaluation of Natural Interfaces to Database Systems". Proceedings of the 19th Annual Meeting of the Association for Computational Linguistics, Stanford.

Some Remarks on:
Language Technology - Myth or Reality?

Dietmar Rösner
FAW Ulm
POB 20 60, D - 89010 Ulm
ROESNER@DULFAW1A.Bitnet

1   State of the Art in Language Technology

In a small sector of the software market, there are products available that deal with written or spoken natural language. Among these software products are machine translation systems, database access systems, spelling checkers, retrieval systems etc. The astonishing fact about most of the products is that they are not based on techniques that are currently under discussion in the field of computational linguistics. This may be due to the necessary time lag between research, development and marketing. In some cases it seems to be a more substantial discrepancy where products are constructed without taking the scientific efforts sincerely into account.

The market for language products is not developed in the same way in Europe and the USA as it is seemingly in Japan. Most, if not all, Japanese computer manufacturers for example have their own machine translation projects. This may indicate that the need for these products is seen differently, or that the estimates about marketability etc. are differing.

In a discussion about progress, one has first to decide about how the progress might be measured. A measure for technologies could be success on the market. It is unnecessary to underline that market success may be unrelated to the academic merits that a linguistic theory might gain.

Instead of answering the question in what application areas today´s state of the art in language technology can produce useful and competitive products, I would prefer to discuss where computational linguistics-based products could and should be developed and marketed.

For a move from academic computational linguistics to "language technology", some preliminary conditions should be met: Lexica and grammars for such products need a substantial coverage of the languages to be dealt with. The software in addition needs to be robust and to a certain degree extensible, and should at best include some learning facilities to account for specialities of applications and for the unavoidable language change.

2   Generic vs Application-Specific Orientation

It is obviously always useful to strive for a generic application-independent technology that allows to easily create specific applications for more restricted cases. This very general motive should be the guideline for all research. On the other hand, it seems feasible to experiment with solutions in restricted areas, even when many questions in general are not yet solved. Humans vary their language behaviour depending on the situation and their specific needs. This gives an additional chance that for certain situations restricted solutions are feasible. An example: As long as recognition of continuous speech is still not sufficiently solved, applications of isolated word recognition should be employed wherever possible (e. g. for menu selection, or in another single command type situation).

The growing maturity of the field may be indicated by the fact as well that there is an emerging awareness of the need for reusable knowledge sources. If one takes into account the amount of human effort that has to be spent for a lexicon with sufficient coverage, it is obvious that reusability of basic knowledge sources like lexica and grammars is a necessary and indispensable condition for the field to become a technology. National enterprises in Japan and the USA and a European initiative have taken this as rationale for their work in electronic dictionaries [AISIG 90]. With respect to grammar, there is no comparable approach yet to be seen.

3   Future Developments in Language Technology

Tools for overcoming language barriers probably have the biggest market potential. The Japanese efforts in MT are - in my opinion - not primarily motivated by scientific challenges, but by the very needs of a highly export-oriented country. Similar needs might be the driving force for language technology work in Europe. The EC is going to launch a programme under the heading of "Language Industry for Europe (LIFE)". In the upcoming common market with the product reliability regulations demanding that consumer products include all necessary documentation in the consumer´s mother tongue, there will be an enormous need for any computational help in solving problems related to multilinguality.

Developments in a variety of subfields of computer science may have strong impacts on language technology, impacts that may be more important than developments in computational linguistics: Advances in hardware (speed, storage capacity, price), parallel processing capacities, database techniques, networking etc.

If one wants to be in the market soon, it is obviously dangerous to take highly complex application areas. As a research strategy on the other hand, it is a recommendable procedure.

4   References

[1] [Coling 88] "Language Engineering: The Real Bottle Neck of Natural Language Processing", panel at Coling 88, Proceedings, pp. 448 - 453, Budapest, 1988
[2] [AISIG 90] Rösner, D.: "Recent Initiatives with Respect To Large Natural Language Systems in West Germany", in: AISIG ´90 Research Workshop "Full-Size Knowledge-Based Systems", Proceedings, George Washington University, Washington, D. C. / USA.

Language Technologies from the Terminologist's Point of View

Christian Galinski
International Information Centre for Terminology
Heinestraße 38, A-1021 Wien

1   Introduction

In general, terminology specialists have little to contribute to the topic "natural language" (that is the concept proper of "general language" with regard to "universal language" including so-called "sublanguages") as far as "general language" is concerned; they can only listen and show their interest. However, as soon as issues of specialized languages with their high amount of terminological units are addressed, they are very well in a position to contribute considerably. This was the actual focus of the critical remarks I made on the occasion of the seminar and which I would like to sum up in this paper. However, I would like to stress that the issues addressed below were not intended to delimitate against each other the disciplines concurring in this field, but rather to find a sensible integration of their theories and methods.

2   The Scope of Terminology Science

It is a given fact that every discipline of science abstracts and reduces its subject proper in order to obtain "scientific" results. On the one hand, this reduction represents a prerequisite for the achievement of scientific results, on the other hand, it might pose problems if it is taken too seriously. Terminologies (i.e. actually "concepts" within their respective "concept system" and their representation) are the scientific objects of investigation of terminology science which can be extended to terminological phraseology as well as to simple statements and systems of statements.

Given the hypothesis that concepts are the smallest units of knowledge, computer-assisted terminography has discovered a variety of applications, pertaining to information and knowledge processing on the epistemological level of concepts, such as, e.g.

- computer-assisted technical writing (concentrating on terminology, terminological phraseology, document management, text management)

- computer-assisted indexing and abstracting,

- computer-assisted conferencing (at subject-specific conferences),

- project-related terminology documentation,

- the function of the conference terminologist, etc.,

with a main emphasis on the knowledge organization and information management aspects rather than on linguistic aspects, with excellent success.

The role of computer-assisted terminography in the field of future "knowledge industries" is becoming increasingly crucial, thus adding an essential component to "language industries".

At their interface with applied linguistics (particularly with research in the field of LSP, language data processing and lexicography) - i.e. with reference to most aspects of the representation of concept with linguistic means, terminology science, terminology work and terminography are open to positive developments. However, linguists and language processing professionals seem to suffer from certain shortcomings, which may be due to communication barriers. Most of them have not, or not sufficiently, kept track of the development in recent years in the field of LSP and terminology research or have not considered them to be relevant. This becomes especially clear when we look at the development of machine translation. In this as in other developments, the lack of integration between terminology science and computer linguistics has led to difficulties in the processing of "specialized languages".

3   Can Specialized Languages be Considered as Sublanguages?

Even if subject specialists did not mind to see his or her specialized language to be placed on the same level as "real" sublanguages such as thieves' cant (which is obviously not the case), scientifically speaking, it is impossible to consider specialized languages as sublanguages. Above all, specialized languages are geared to specific subject fields and their development is not primarily subject to language-immanent tendencies.

Terminology science differentiates methodologically and epistemologically between

a) general language, with an, in principle, uncontrollable number of linguistic phenomena, and

b) specialized languages, for which a control of the number of linguistic phenomena is largely desirable (particularly as far as natural sciences and technology, i.e. prescriptive subject fields, are concerned) or at least part of the methods of knowledge representation (as far as the humanities and social sciences are concerned) would be appropriate.

The intersection with terminology science would increase considerably if specialized languages, from the point of view of text linguistics, were to include non-linguistic representations.

4   Are Terminologies Part of the Lexicon of Universal Language?

Concepts are created in order to render manageable the multitude of charactaristics of (material or abstract) objects. Concepts themselves are also abstract objects. In order to record, represent and process them, the following features are needed:

1) *symbolic representation* with linguistic symbols (facilitating oral communication) and/or non-linguistic symbols (increasing the efficiency of written specialized communication),

2) *descriptive representation* with linguistic symbols (such as definitions, explanations, co(n)text, etc.) and/or non-linguistic symbols (such as complex formulae, graphs, diagrams, etc.),

3) relations between individual systematic data categories and individual data fields (ad hoc) within a single concept entry and between concept entries of equal (logical files) or even different files, in order to establish the "System of Knowledge", however evolutionary and unsystematic it may be,

4) *macro structures* (such as classifications, documentation thesauri and other documentation languages).

Thus, the terminologies as described above are not, and will never be, totally integrated into universal language.

5   Terminological Praseology vs. Collocations

At the occasion of a workshop on terminological praseology at the end of 1990 it became clear that

- there are multi-word units (e.g. noun-verb combinations) which have traditionally been regarded as phrases, but, since they represent concepts, should be included in the category of multi-word terms,

- there are symbolic representations of concepts (particularly in the field of technology), which represent a combination of linguistic and non-linguistic symbols,

- it is sensible (at least for methodological reasons) to draw a clear-cut line between LSP phraseology and collocations.

A methodological distinction is necessary when it comes to analysing and processing specialized texts, but it certainly implies subject knowledge and an understanding of the conventions of technical writing in a particular subject field.

6   The Development of Specialized Languages and Terminologies

Terminologies of individual subject fields develop in the wake of scientific technical and economic industrial progress. Their aim is to represent specialized knowledge (also in specialized communication). The terminology of steel of an individual country, for instance, depends above all on the composition of the raw material and the specific characteristics of the steel of this country as well as on the methods used to produce and process steel. It is not primarily a feature of language development.

As a result of the proliferation of knowledge and due to the fact that science and technology are very much sectorised, the specialized languages of different subject fields are developing in different ways and with different dynamics. In specialized communication, terminologies are embedded in specialized languages. For decades now, the development of many languages has been influenced by the development of specialized languages. Thus, it is not the "créativité de la langue" (creativity of language) that is primarily responsible for language development, but the creativity of scientists and other subject specialists.

Due to the limited number of term elements that can be used to designate new concepts, every language (including English) faces difficulties to provide a sufficient number of unambiguous terms for communication in specialized languages. Subject specialists, who are the main "inventors" and users of "their" respective specialized language, are at the same time also creating linguistic problems (in the form of homonymy, synonymy and quasi-synonymy). They are then trying to solve these problems via terminology work and particularly through terminology standardization.

7 Specialized Language and Documentation

A major part of human specialized knowledge is laid down in writing, i.e. as (printed or electronic) documents, by means of specialized languages. Of course, not all human knowledge can be laid down in written form; some kinds of knowledge cannot be represented in writing at all and, therefore, cannot be processed by a computer.

Increasingly, documents written in a particular specialized language (i.e. texts) contain non-linguistic forms of knowledge representation. Epistemologically, this can be explained by the fact that language itself is subject to certain limitations and, thus, has to be complemented by non-linguistic representations. Today, the "written" text is probably the most widely used means of communication among subject specialists. The most important knowledge units in these texts are, above all, terminological units and their terms (which include an increasing number of non-linguistic symbols).

The above illustrates that terminology science is closely related to the philosophy of knowledge and epistemology as well as to information science.

The subject-related application of terminology science with its terminological methods and data is part of the indispensable and essential equipment of each and every subject field. Unfortunately, this fact is very often neglected in higher education.

8 The State-of-the-Art of Terminology Science and the Methodology of Terminology Work and Terminography

Concepts considered in the framework of terminology science and terminology work are:

- units of thought for recognizing "objects" as part of reality,

- units of knowledge for ordering objects and related knowledge,

- units of communication facilitating the representation and transfer of knowledge from one subject field to another.

Concepts never occur in isolation; they are linked to each other through different kinds of relationships. Thus, terminology science is concerned with:

- concepts and the relationships between them,

- concept representation by means of:

    - representation through symbols, such as

        - terms,

        - graphic symbols,

        - combinations of either,

    - descriptive representation, such as for instance

        - definitions, explanations, etc.

        - (complex) graphic representations, illustrations, formulae and other forms of non-linguistic representation,

        - combinations of concept descriptions and non-linguistic representations,

    - the structure of material and abstract objects,

- the natural and mental order of

    - objects,

    - concepts,

    - concept representations.

Terminography, particularly computerized terminography, offers practical methods of recording and processing terminological data. Terminography is closely related to information science, especially to Information and Documentation. Today, it is safe to say that information (resources) management without appropriate terminological methods and data is no longer imaginable or, at least, will never be efficient.

This tremendously influences our understanding of "text", of what we can do with it and *how* to go about it.

With respect to specialized texts, today, every kind of symbolic representation of information and knowledge in modern text linguistics has to be considered as text, including non-linguistic elements. Since units of text (e.g. according to SGML philosophy) can again represent identifiable texts, such a text can, in the extreme, consist of:

- one symbol only,

- non-linguistic representations only.

9   Terminology & Documentation - T&D

Efficient Information and Documentation (I&D) cannot do without terminological methods and data. Terminology work and terminography cannot do without I&D methods, if its aim is to produce high quality and reliable data. That is why Infoterm attempts an integration of the theoretical and practical methods of terminology work with those of information science and documentation via "Terminology & Documentation" (T&D). In this context, terminologies represent the microstructure of knowledge, whereas documentation languages (e.g. classification systems, thesauri, etc.) represent the macrostructure of knowledge.

This integrative approach towards T&D has led to an unexpected increase of efficency in various applications as exemplified by the method of computer-assisted conferencing developed and carried out by TermNet, the International Network for Terminology.

10   Applications of T&D

Terminological methods and data and their integration into T&D with appropriate computer support offers a variety of applications, e.g. in

- technical writing (by subject specialists in general or by technical writers, marketing experts, etc.),

- information processing and information retrieval,

- specialized language teaching,

- teaching and training of/in a specific subject field,

- knowledge and technology transfer, etc.

In addition, a variety of computerized or computer-assisted procedures for processing subject-related data can be applied to fields such as machine or computer-assisted translation.

Repeatedly, knowledge acquisition by students in a particular subject field has proved to be less time-consuming, when the subject was presented

- systematically, i.e. in a clear order reflecting the knowledge of a specific subject field (requiring adequate terminology);

- unambiguously in a given context, i.e. by means of unambiguous terms, without numerous stylistic variations.

In this respect, the applications of terminological methods have proved to enhance the performance of the students (and, by the same token, of the teachers).

Terminological methods and data have helped to design most efficient didactic methods in subject-related foreign-language courses (such as one-week intensive Russian language courses for engineers, four-week intensive Japanese language courses for architects) which have been carried out successfully. The main aim of such a training is to teach the ability to understand, analyse and evaluate specialized texts in order to make appropriate use of the information inherent in printed or electronically stored specialized texts.

The advantages offered by the application of systematic T&D methods to an individual country are obvious.

11 Conclusions

1) Due to the language-immanent limitations of the efficiency of language in view of knowledge representation, specialized knowledge can never be solely represented by linguistic means (nor possibly by other "text means").

2) Due to the complexity of several aspects of knowledge representation and knowledge processing, no single discipline should try to monopolize this domaine. Only an integrative approach will be of any help, with terminological methods and data being indispensable.

3) The so-called "information crisis" is not primarily a linguistic or quantitative problem, it is rather a problem of lacking consistency and order within the amount of information.

The Development of Short and Medium Term Language Industries

H. Schnelle
Ruhr-Universität Bochum
Inst. für Allg. Sprachwissenschaft
Universitätsstraße 150, D-44801 Bochum

1   Introduction

As linguists, we know our problems in syntax, morphology or semantics and we try to solve them with the best scientific methods available. We also know, approximatively, what would count as a scientific solution. Scientific solutions are the necessary prerequisite for the development of language industries, but their availability is not sufficient at all. In this sense research in pure linguistics is only partially relevant for the development of language industries.

Stuck to our problems, we are even surprised or shocked by the conception of a language industry. Is there such a thing? It is necessary to step back from our everyday work and to see the practical and social relevance of what we are doing in historical perspective.

Originally, people only had speech and gestures. The first step towards technology was to manufacture pictures and pictorial symbols on walls and on stone and, from these, more abstract symbols leading to phonetic writing. The second step was the production of appropriate storage material for passive storage - parchment, paper etc. - on which storage and transmission of symbols was much easier. The third step was the mechanization of the production process - print - which allowed an enormous multiplication of copies of written language products.

We are presently involved in the historic transformations connected with the fourth step: Storage of language products on active material, namely on computer-readable media. Even if this did only involve the task of copying the wealth of accumulated encyclopedic knowledge, literature etc. in order to enable everyone to consult or read it on his or her computer terminal, it would constitute an enormous amount of work for an industry transforming printed material into computer-readable copies. The task would be similar but infinitely larger than that of the Gutenberg era. However, though most of the material to be manipulated would be language material, we would not be tempted to call the industry which would be in charge of this a "language" industry.

The situation is, however, radically changed by the fact that the language material will be stored in an active medium whose activity can be made use of and systematically controlled by program. Computers are in fact primarily symbol processors rather than number crunchers. Thus the processes of writing and understanding of texts are - in principle - natural tasks for them. Implementing the details of what is thus possible in principle requires to provide the computer with machine readable dictionaries and grammars to program. This would lead to processes in the machine which should be similar to the processes in humans when they read a foreign language with the help of grammars and dictionaries. However, machine readability of ordinary dictionaries and grammars is not sufficient. Instead, dictionaries and grammars must first be coded to enable the programs to interprete them.

This was the first idea of computational linguistics. Unfortunately, it became quite clear that the understanding of a text not only depends on dictionaries and grammars but also on various contextual factors, such as common sense knowledge as well as the understanding of the particular situation in which a text is uttered. In principle these different types of knowledges can also be represented in and interpreted by the computer.

It thus turns out that implementing the simple idea of mechanization of language understanding becomes a rather complicated task. Computational linguists have been concerned with it for fourty years and still are. Though the difficulties are enormous, there is no indication whatsoever that they can not be overcome for typical uses of language understanding. There cannot be any doubt, however, that the production of language competence in the computer will require the creation of a new technological discipline, language technology, and will result in an industry of products which provide language competence to the computer and to automatically controlled products - from grammar and style checkers to translators and from the "speaking" automobile over more general information systems to robots and automatic factories. This is, however, a long term task.

What was said seemed to imply that there are two radically different problem areas: the transcription of written language data into electronic data and the construction of language competence for computers and automatic systems. There is, however, an area of intermediate problems: these concern the currently available printed grammars and dictionaries. It is assumed that they stand in sharp contrast with grammars and dictionaries coded for automatic use. Whereas the latter require the explicit representation of every detail of knowledge the former are not fully explicit and partially unsystematic but still optimally adapted to the needs of human users. It seems that for the consultation on computers, say in connection with a word processor, they do not need reworking but merely transformation into machine readable form.

This last evaluation does not stand against closer analysis. Existing dictionaries are not optimal and it is possible to develop better dictionaries and grammars for human users on the basis of detailed insight into understanding processes acquired in the last decades. The development of better dictionaries and grammars for human users is not in contrast to the construction of dictionaries for computational use. Rather, the best dictionaries for computational use should comprise all the information necessary, from which excellent human dictionaries and grammars could be derived. The relevant data for human use should be produced by eliminating the information of the data base which is necessary for the computer but obvious and trivial for a human user and would only disturb him when consulting the dictionary or the grammar. We thus see that the immediate task of the language industry is to develop, to test, and to evaluate better knowledge representations (grammars, dictionaries, terminologies) supporting human users (e.g. writers, language learners, translators, etc.) parallel to the rather long range task of developing fully automatic language competence in automatic systems.

The development of language industry - transcription of language data, improvement of computer accessible representations of the knowledge of language and the construction of language competence in automatic systems - must be systematically organized. In this organization the following *dimensions* must be considered:

1. factors,

2. criteria,

3. strategies,

which determine or characterize this development. We shall now discuss them in turn.

2   Factors

We are involved in a process which has already happened in many other areas: the transformation of the arts and crafts into a technology. The transformation of writing into printing was the first step, the reasoned styling and transforming of written or spoken texts within a language or between languages is the second. The transformation of the tools which are appropriate for such styling and transforming, i.e. the adaptation of existing grammars, dictionaries, elements of rhetoric, etc., to the principles of rigorously reasoned processing, constitutes the initial step of the general process of automatization whose subsequent steps are the automatizations of reasoned language processing.

The details of this process must be determined within linguistics and within the linguistic technology to be developed. The general form of the process is, however, the same as in any process of transforming the arts and crafts of a domain into an industry.

The *factors of transformation* are: transformation of

2.a.   knowledge into science,

2.b.   skill into technology,

2.c.   immediate marketing into economic planning and marketing of products,

2.d.   immediate social role into sociologically controlled adaptation to the general scope of human needs and values.

These factors must be characterized with respect to the citeria and strategies of scientific, technological and production development. We shall first turn to the criteria.

3   Criteria

The criteria are different on each of the following three levels of

3.a.   principled insight,

3.b.   insight into the conditions of feasibility, and

3.c.   insight into the organization of production.

The last decades have been mostly concerned with the first factor on the first level: the transformation of informal linguistic insight into principled scientific knowledge and detailed description. The transformation of the knowledge of the form of languages is presently rather advanced. It turned out that the analysis of the details were much more complicated than one first thought. In contrast, the knowledge about the meaning specifications for languages is still meager. Some principles are known which are applicable to constructed languages; they have been tested on fragments of ordinary languages. Several decades of research will still be necessary before insights in semantics can match present insights into the forms of languages.

Indeed, the theory of the linguistic form is rather advanced (though far from complete): We have strict criteria of the form of description, of their empirical adequacy, as well as of their consistency. This holds even for the details of description concerning some languages such as English. The same cannot be said for the meaning (semantics), still less for the principles and details of language use, such as their appropriateness in contexts of communicative situation and presupposed world knowledge (pragmatics).

So much for the linguistic means, i.e. the statics of the forms used. As to the dynamics, i.e. the description of the processes which implement the production or detection of form in the user, computer models of sequential and rule controlled use are rather advanced whereas the analysis of massively parallel procedures, as they occur in human beings, is not yet well understood.

The best computational models of the dynamics of linguistic form analysis have reached a stage of laboratory models from which technological products could be developed. They even allow the evaluation of some of the *construction criteria of technological products:*

- feasibility,

- efficiency,

- robustness.

Given this evaluation of the state of the art of pure linguistic science and applied computational linguistics, the central question for short and medium term technological development is the following: Is the automatization of partial linguistic competence (linguistic form with the exclusion of understanding) sufficient to generate useful products? The answer must take current experience into account: Though the quality of automatic translation produced by current systems is still low, it is surprisingly high if one considers that the systems do not have any understanding of the meanings of what they actually translate. This indicates that the scientific insight into the details of the functioning of the languages may be sufficient for the construction of useful technological products.

We thus might proceed to the practical criteria. In particular, we now need a clearer picture of the *economic and social criteria* of technological development, i.e. of

- Demand,

- Cost (perhaps: feasibility of cost-reduction),

- Organization of the production process,

- Benefits of the products,

- Social consequences,

- Subsequent costs.

The analysis of these criteria must be evaluated differently with respect to the different strategies of development, that could be adopted.

4   Strategies

I shall distinguish two strategies of development: revolutionary and evolutionary. According to the first strategy, there does not seem to be much correspondence between existing procedures and products and the features of technological products to be constructed. Its perspective of a language industry is that of an industry to be created from scratch. Revolutionary strategy insists on a new construction in very detail. It assumes that the technological products are radically different from the means and tools of the arts and crafts it intends to replace. The development of aeroplanes is a case in point. They are different from flying animals in almost every aspect.

In contrast, the evolutionary strategy is adaptive and piecemeal: Existing skills are studied and reconstructed and existing tools are reorganized and rigorously controlled in view of their integration into gradually transformed increasingly technological products and processes. This strategy does not exclude the revolutionary development of certain components. These are, however, integrated into existing frameworks.

The attitude taken by scientific research is usually revolutionary. This is as it should be. The processes and the products to be constructed should be understood and constructed in systematic principle and in every detail without presupposing features which seemingly do not require explanation. In contrast, the technological attitude is practical and usually adaptive. A typical case is the automobile. A component, the combustion engine, was invented. In view of its practical use, it was put on an only slightly adapted horse car to build the first automobile. It was by piecemeal adaptation that, step by step, reliable and useful cars were developed.

The conception of the automatization of language competence has been mostly revolutionary during the last decades: Construction of systems for machine translation should be based on radically new conceptions of grammars and dictionaries. The resulting products, such as systems for machine translation, should be based on grammars and dictionaries which are totally different from ordinary ones. The approach has been successful in contributing to a fruitful development of pure linguistics and applied computational linguistics as outlined above. Given the still underdeveloped state of semantics and pragmatics, i.e. of the science of language understanding, it is, however, doubtful whether the revolutionary strategy could soon lead to useful products. It is advisable to reanalyse the situation in the light of the evolutionary strategy.

In the evolutionary perspective, language industry exists already and it is our task to improve its products gradually. It is constituted by those institutions which produce and sell language products on a market. Existing language industry is almost completely precomputational. It comprises production industry such as publishing houses of dictionaries and grammars and services such as translation bureaus or bureaus for generating and applying terminologies. It also comprises some software houses which produce word-processors, spelling checkers, style checkers etc. Due to support for research - either public support such as Esprit-projects etc. or inhouse support such as in the case of the LILOG project of IBM-Germany - some precompetitive laboratory models for more complicated products have been developed.

In spite of these newly emerging sections of the language industry, it is not computer usable products which sell; for example, the users of the dictionaries which sell are human end-users, namely learners of a foreign language or those aiming for a better command of their own language. In the activity of these people language products do play a role as tools helping them to do other things. Since the products sell, it is usually assumed that they are well adapted to their purpose. Improvements of dictionaries do not seem to require better structure but only the addition of recent words.

This is, however, a fundamental misconception. Existing language products in general and dictionaries and grammars in particular are far from optimal with respect to the relevant information they provide. This insight is not obvious, though. The reason is that the quality of the products is difficult to evaluate, because the users are intelligent beings which can extract information from sources which are rather poorly specified, as is the case in ordinary dictionaries. Dictionaries and grammars for human users could be radically improved in their organization and perspicuity. This is not to say, that they are wrong or really bad in their present form. Quite to the contrary, large dictionaries contain a wealth of information, which, in spite of their being rather unsystematically organized cannot be dismissed in the development of computer-usable products, such as machine translation systems or natural language components for machines and systems that can be addressed or controlled in ordinary language and which can react in ordinary language.

These considerations lead to the conclusion that the evolutionary strategy of the development of language industry should be applied in parallel to the still dominent revolutionary strategy of the development of radically new products of the automatization of language competence.

In the spirit of the evolutionary strategy, we should think about ways of stepwise improvement of existing products, in particular of dictionaries and terminologies. These products should be as complete in their information as possible and required for the specification of the automatization of language processes. At the same time they should be provided with filters which suppress information that is obvious for a human user as soon as a human user consults the data base.

However, it must be clearly understood that the improvement of dictionaries and terminologies, which are by far the largest subcomponents of language understanding systems, requires a high degree of organization. It can only be guaranteed by computational tools which on the one hand provide quick access to any information that may be relevant in the lexicographers decisions but on the other hand control the consistency of these decisions and the completion of necessary cross-reference.

Linguistic research in computational linguistics following the revolutionary strategy has largely concentrated on the development of grammars in the last decades and has made big progress in theoretical understanding. This progress was the result of very concentrated and highly sophisticated research based on principles of mathematical formalization of symbol manipulation. But the real challenge will come with the task of integrating large dictionaries (at least 200.000 readings of words) with sophisticated kinds of grammars and with the attempt to derive an understanding of the meaning from the grammatically and lexically structured expressions.

Recent linguistic analysis has shown that the meaning contained in a dictionary can be expressed by the inferential and paraphrastic relations between words and sentences with which word explanations are given in a dictionary. In fact, already the system of word explanations or definitions given in good dictionaries published in recent years can be understood as inferential knowledge bases in the technical sense of artificial intelligence research. It is here that language industry is confronted with a real and technologically interesting challenge: to improve existing dictionaries by organized work of lexicographers aided and consistency-controlled by lexicographer's tools to be developed in the years to come. The result should simultaneously produce better paper dictionaries of the ordinary type and computer-usable dictionaries which could serve as components in medium and long range products of natural language systems.

5   Summary

The development of short and medium term language industries requires a re-evaluation of computational linguistics. As a long range task implementing language competence on computers according to rigorous principles of scientific linguistic research remains important. The structure of the resulting products, such as machine usable dictionaries, grammars, representations of knowledge and representations of descriptions is seen as fundamentally different from corresponding representations for human users, which are considered to be already optimally adapted to their task. It is here that a revision of perspective is needed. Current dictionaries and grammars are not optimal for human users. They need more systematic organization and even reorganization, without overloading the user with irrelevant information. This reorganization is only possible by using computer tools which operate on machine-readable transforms of existing dictionaries and grammars, providing quick access to comparisons, and checking consistency and completeness of entries. With the help of these tools all information about words and constructions that is relevant for humans or computers should be gradually introduced. Information that is trivial for a human user should then be filtered out whenever such a user has access to the machine, or whenever a printout for a conventional dictionary is produced. A similar strategy could be followed in the development of grammars.

The strategy is evolutionary. It immediately produces products, which are better than the existing products and, at the same time, are gradually transformed into a system that is not only computer-readable but also computer-usable by the programs which implement language competence. Typically then, the evolutionary strategy is the one adapted to short and medium term goals.